OCR

Revise Geography

A2

...e better support for you

Chris Martin

Official Publisher Partnership

Heinemann is an imprint of Pearson Education Limited, a company incorporated in England and Wales, having its registered office at Edinburgh Gate, Harlow, Essex, CM20 2JE. Registered company number: 872828

www.heinemann.co.uk

Heinemann is a registered trademark of Pearson Education Limited

First published 2009

13 12 11 10
10 9 8 7 6 5 4 3 2

British Library Cataloguing in Publication Data
A catalogue record for this book is available from the British Library.

ISBN 978 0435 357 71 9

Edited by Kelly Davis/White-Thomson Publishing
Designed by Kamae Design
Typeset by Phoenix Photosetting, Chatham, Kent
Original illustrations © Pearson Education 2009
Illustrated by Phoenix Photosetting, Chatham, Kent
Cover design by Pearson Education
Cover photo/illustration © Superstudio/Getty Images
Printed in Malaysia, CTP-KHL

Acknowledgements
The author and publisher would like to thank the following individuals and organisations for permission to reproduce copyright material.
Page 34: Figure from Guinness, P. and Nagle, G. *AS Geography Concepts and Cases* (2000), Hodder Murray, reprinted by permission of John Murray (Publishers) Ltd.
Page 53: Table adapted from *New Scientist* magazine, 26 May 2007, pp. 34–41.
Page 67: Table adapted from *UN World Economic and Social Survey 2006*, Maddison (2001) and UN/DESA, reproduced by permission of the UN.
Every effort has been made to contact copyright holders of material reproduced in this book. Any omissions will be rectified in subsequent printings if notice is given to the publishers.

Websites
There are links to relevant websites in this book. In order to ensure that the links are up to date, that the links work, and that the sites are not inadvertently linked to sites that could be considered offensive, we have made the links available on the Heinemann website at www.heinemann.co.uk/hotlinks. When you access the site, the express code is 7719P.

Contents

Introduction iv
Chapter 1 Earth hazards 1
Chapter 1 Exam café 13
Chapter 2 Ecosystems and environments under threat 15
Chapter 2 Exam café 26
Chapter 3 Climatic hazards 28
Chapter 3 Exam café 45
Chapter 4 Population and resources 47
Chapter 4 Exam café 62
Chapter 5 Globalisation 64
Chapter 5 Exam café 76
Chapter 6 Development and inequalities 78
Chapter 6 Exam café 94
Chapter 7 Geographical skills 96
Chapter 7 Exam café 107
Conclusion 110
Exam café 112
Answers to quick check questions 114

Introduction

This book is designed to aid your A2 Geography revision. It covers all the major points in each topic as they appear in the OCR specification. You can fill in the details from what you have been taught or from the Heinemann student book written to support the specification (this book contains numerous cross-references to the relevant pages in the student book). Much of the material in this revision guide is an enlarged version of the 'Refresh your memory' sections in the student book and on the student support CD-Rom. The revision tips and other advice are drawn partly, but not exclusively, from the student book, and this shortened version contains all the most important elements in the course.

How does A2 differ from AS?

At A2 there is less emphasis on facts and more on analysis and evaluation. There are fewer 'right' answers, and a logical opinion supported by examples will always be well rewarded. Content is less important than thoughtful reflection. There is more emphasis on essay writing; and essays are the main way in which candidates can reach the A* grade, which is only available at A2.

There are two key differences between AS and A2. At A2 the key skill is that of being able to evaluate, so all essay questions require candidates to weigh up the pros and cons of a situation. Also case studies are expected to be more detailed at A2. A simple 'e.g. Birmingham' is totally inadequate at this level.

Revision tips

How do you learn everything you need to remember for the examination? First you need to work out what sort of learner you are – which learning methods you prefer and what time of day suits you best.

What learning style are you?

You can discover your learning style by looking at the table below. Which do you find the easiest way to learn and recall information?

Learning style	Preferred revision methods
Visual	Drawing maps and diagrams Using spider diagrams Using headings Using mnemonics Using a CD-Rom or the Internet
Auditory	Recording someone reading your notes Making up rhymes or songs Getting someone to ask you questions
Kinesthetic	Cutting up sections of your notes and rearranging them Building models of things Using acting and role play Compiling a card index

Most geographers are visual learners so the student book makes use of mnemonics (pronounced '*neh-mon-iks*'), where a word or phrase is used to recall a list of information, and spider diagrams.

Mnemonics

The mnemonics used as memory aids in the student book include:

- PESP (geographical factors) = Physical, Economic, Social and Political
- CRDVDS (physical or environmental factors) = Climate, Relief, Drainage, Vegetation, Disease (and wildlife), Soils (and geology)
- SPITMAS (human factors) = Settlement, Power, Industry, Transport, Mining (and quarrying), Agriculture, Services

Spider diagrams

These diagrams make ideal essay plans because they show the interconnections and logic of geographical factors and features.

Timing

Remember that some people can revise more easily at certain times of the day. Work out when you learn best – morning, afternoon or evening. You should also remember to take regular breaks (10–15 minutes off in every hour) and don't revise all day. Revise in manageable chunks or topics. You will need to go over each chunk several times before it stays in your long-term memory.

The specification

If you don't have a copy of the specification, check it out on the OCR website (go to www.heinemann.co.uk/hotlinks, enter the express code 7719P and click on the relevant link).

At A2 there are two papers: Global issues (worth 30 per cent of the A level) and Geographical skills (worth 20 per cent of the A level).

Global issues

Global issues looks at contemporary issues at global, regional and local levels.

There are six topics to choose from:

A. Environmental
  1. Earth hazards – earthquakes, volcanoes, mass movements and floods
  2. Ecosystems and environments under threat
  3. Climatic hazards – hurricanes, high and low pressure systems and human impact on climate

B. Economic
  4. Population and resources – population growth compared with the resource base
  5. Globalisation – its causes and impacts, including trade and aid
  6. Development and inequalities – the causes of economic, social and environmental inequalities

You should have studied three of these topics, selecting at least one from each group.

The material is set out under three headings: Questions for investigation, Key ideas and Content.

Questions for investigation

These questions set the broad context and pose questions that could be answered through research and fieldwork.

Key ideas

These key ideas present some of the approaches that could be used to answer exam questions. Essay questions tend to ask you to evaluate aspects of the key ideas.

Content

This gives a list of essential topics and concepts that you can be expected to cover in the time available. You can go beyond this 'bare minimum' list but don't try to leave any out. This is the material that you should select from to support your discussion.

Each topic has five, increasingly complex questions to investigate. The first section is usually quite factual and is related to patterns and processes. By section five, the questions are more complex and open to debate, for example focusing on efforts to make the system more manageable or sustainable.

Geographical skills

The Geographical skills paper covers the research skills you need to carry out an effective investigation. The content is arranged under the six stages in the research process:

1. Formulating a question for investigation
2. Planning the investigation
3. Collecting the data
4. Presenting the data
5. Analysing the data
6. Evaluating the investigation.

Case studies and examples

Throughout the examination you are expected to refer to real places and locations in your answers even if this is not specifically asked for in the question. The specification sets out the minimum number of examples or case studies needed to allow an evaluation of a particular topic. These should include contrasts in some aspects, which then allow you to make comparisons.

At A2 more detail is expected and it is better to know a few examples in depth rather than many superficial ones. Sketch maps to show locations can be useful but may sometimes be time-consuming distractions.

A local small-scale example from your own area is often quite powerful, and examiners like to see examples that go beyond the ones given in the student book. Some new case studies have been added in this revision guide.

Assessment objectives

The exam papers have been set out to match the three assessment objectives and these have a different weighting than at AS level.

AO1 – Know and understand concepts, processes and facts

This objective accounts for only 33 per cent of Global issues and 25 per cent of Geographical skills.

AO2 – Analyse, interpret and evaluate issues and apply understanding

This objective is worth 50 per cent in Global issues and 25 per cent in Geographical skills, and tends to be assessed in the essays.

AO3 – Investigate, conclude and communicate

This objective accounts for 50 per cent of the marks on the Geographical skills paper and 17 per cent in Global issues.

The examination

The two papers are very different. Global issues is 2.5 hours long and is marked out of 90 (60 marks for the two essays). Geographical skills is 1.5 hours and is marked out of 60 (with 40 for the two essays).

Each paper has two sections.

Global issues

Section A consists of short 10-mark questions designed to get your brain functioning. There is one question per topic and you must answer three, at least one from each section – environmental and economic. **They do not need to be from the three units you studied.** You can use material from your AS and wider reading to answer questions. Each question covers a particular resource and you are asked to identify one issue and suggest appropriate strategies to manage it. The emphasis is on the strategies so don't waste time hunting for hidden issues. You should spend no more than 20 minutes (including reading time) on each of your three answers.

Section B gives you a choice of one out of six environmental questions (two per topic) and one out of six economic questions (again two per topic). You should aim to spend 45 minutes per essay. **They all require you to support your discussion with specific located examples.**

Geographical skills

Section A has short questions in three parts, with the first two sections (usually 5 and 10 marks) based on a resource taken from an investigation. The third part (worth 5 marks) opens up the question to broader research issues.

There is a choice and in any exam you can expect questions based on:

- A physical geography investigation
- A human geography investigation
- A generic issue in investigations.

You are required to answer one question and it should take you about 30 minutes.

Section B is based on your own investigations at AS and A2 and requires you to evaluate aspects of the six stages of your investigation. There are two compulsory questions. You should always try to personalise these answers, using your own experience, and you are required to state the title of your investigation.

If you run out of space you can use extra sheets. Both papers will be marked online so try to avoid using colours on diagrams.

Question wording

It is vital to understand command words in the questions, as they will direct your approach. Some terms are regularly misunderstood by candidates.

Terms that are often used in the A2 Geographical skills paper include:

Accuracy – conforming precisely to a measurable standard (usually at least 95 per cent)

Analysis – examination of the components of a whole; investigating how the separate elements fit together

Correlation – a relationship (whether positive or negative) between two variables

Evaluation – determining the value or importance of something

Hypothesis – a provisional assumption made in order to investigate its consequences

Interpretation – explaining what the evidence means

Level of confidence – the extent to which we can be sure that the result is valid and secure (normally set at 95 per cent or better that the null hypothesis was not proven)

Null hypothesis – when the result/correlation came about by chance

Reliability – dependability; the extent to which a piece of research will produce the same results every time it is repeated

Representative sample – a sample that is large enough to reflect fairly the nature of the whole.

Command words commonly used in both Global issues and Geographical skills papers include:

Justify – prove or show why; provide evidence on which to base your choice of one thing rather than something else

Discuss – debate a topic or issue; present the points for and against

Examine – investigate closely; describe, explain and comment on

To what extent – assess; evaluate.

Question types

These three types of question each require a different approach.

1) Data response (Section A in Global issues and Section A in Geographical skills)

This is usually the first part of a question and it is designed to see if you can interpret (read or understand) a diagram. Sometimes you are expected to describe a trend or state the value of something. You can read the values from a graph or diagram, using the key or the scale on one of the axes. You must quote actual values. Saying 'it's high' isn't good enough. Trends can be positive or negative, strong or weak. Always look for odd values (anomalies) that don't fit the general pattern.

2) Short answers (Geographical skills paper only)

These questions are worth either 5 or 10 marks each, and they normally ask you to explain something related to the data response. They test your knowledge and understanding – for example by using alternative techniques to the one shown. These questions should be developed in depth. **Remember: depth is more important than a lot of superficial points.** This is where your mnemonics will come in useful. You can also use appropriate diagrams and they may reduce the amount you have to write. For instance, it is difficult and time-consuming to describe a scatter graph. These questions are marked on levels – the higher levels require clear explanation and the correct use of geographical terms.

3) Essays

Essays are worth two-thirds of the A2 level marks and are designed to reveal those candidates who can construct effective arguments. These are the types of challenging questions that positively encourage you to 'think outside the box'. There is space on the paper for you to plan your answer. **Remember: if you do run out of time to finish your answer then this plan will be looked at to see what you had intended to write**. Planning is important to structure your answer. Each paragraph should have a distinctive grouping of ideas, e.g. Physical, Economic and then Social.

Your introduction should show your understanding of the question, the line of argument you will follow and the examples you will be using to support this argument. The conclusion of your essay is equally important. This summarises your argument and draws together what has gone before. Agreeing or disagreeing with the initial question is less important than giving the chief reasons why. There is rarely a totally definitive answer, especially in Global issues essays. It usually depends on where you are (e.g. rural versus urban, coastal versus inland, developed country versus developing country) and the group's viewpoint (e.g. rich versus poor, young versus old, single versus family).

Try to understand the paper

Always get hold of past papers and use them to find out how the examiner words the questions. You can use these to practise reading and answering questions within the time allocated. **(But beware – there is a risk you will put in your exam paper what you hope the examiner is asking or what they asked previously.)** You can also look for patterns of questions. All exam papers are set using a specification grid to help avoid duplication or repetition. January and June examinations are set at the same time – usually two years in advance. What was in the news then? Even better, get hold of the examiner's report, on the exam board website or from your teacher, as this contains helpful advice on how to improve performance, paper by paper and question by question.

The layout of this revision guide

This revision guide follows the headings and content in the OCR specification and the Heinemann student book, but it also includes some new content to enhance case study material.

Key words

These are some of the geographical terms that you will be expected to know and use. They are defined in the student book.

Diagrams

Diagrams are included for some of the topics. These are not meant to be definitive. Rather they indicate ones that you might find useful in your answers to illustrate major concepts. Remember they do not have to be great works of art, as they need to show a geographical concept in a clear way but in limited time.

Quick check questions

These are designed to test your knowledge of some of the revision points at the end of each section. These are not the factual types of questions you can expect in the examination.

Exam tips

These are helpful general suggestions relevant to the OCR examination rather than hints about actual questions.

Exam cafés

These occur at the end of each chapter and are based on the style of questions you can expect in the examination. In some cases typical diagrams that you might encounter in the examination have been included and in some cases they have been omitted to save space, but their nature should be clear from the text and comments.

Chapter 1 Earth hazards

Student book pages 6–45

1.1 What are the hazards associated with mass movement and slope failure?

Student book pages 8–15

Causes

- Reduction in shear strength and increase in shear stress
- Slope system and angle no longer in equilibrium so needs to change angle (dynamic equilibrium).

Factors holding slopes in place

- Friction and weight of particles
- Cohesive forces, e.g. clay
- Vegetation roots bind the soil
- Human structures, e.g. nets.

Ways of classifying mass movements

Student book page 10

- By speed or rate of movement – fast versus slow
- By type of movement – e.g. slide versus flow versus creep
- By type of material – rock versus soil versus ice
- By water content – wet versus dry.

Processes

- Creep – slow and low water content results in terracettes
- Avalanche – rapid can be wet or dry
- Flow – rapid highly fluid, saturated
- Slide – sliding material retains shape and cohesion
- Slips – on slide plane, medium water content
- Slump – usually rotates along a slip plane, medium water content.

Key words

Shear strength	Shear stress
Slope failure	Dynamic equilibrium

Factors increasing mass movement		RS/SS
SS = increases shear stress RS = reduction in shear strength		
Physical	Climate – wet, lots of weathering, extremes of temperature	RS
	Slope angle – steepness of slope	RS
	Drainage – wet areas are lubricated	RS
	Rock type – geology, e.g. clay, structure, beds, porosity and tilt of rocks	RS
	Vegetation – type and percentage of cover	RS
	Animals – burrowing animals, walking on slopes	RS
	Erosion of foot of slope, e.g. waves	SS
	Sudden shocks – earthquakes	SS
Human	Over-steepening slopes, e.g. cuttings, quarries, adding waste	SS
	Removing vegetation, especially trees	RS
	Drainage – adding moisture to ground adds weight and lubricates	RS
	Adding weight to slopes, e.g. buildings, walking	SS

Impacts of mass movements

Environmental impacts

- Relief – reduces slope angles, fills in valleys/hollows, adds steps or bulges to slopes
- Drainage – may dam or divert rivers, wetter at foot of slope
- Vegetation – trees lean or fall
- Soil – collects at base of slope (catena effect)
- Rock strata – may bend the ends of beds (cambering).

Social impacts

- Buildings/walls – collapse, lean or have soil collect up side of slope
- Disasters, e.g. Aberfan 1966, 147 killed.

Economic impacts

- Transport – road and rail distorted or broken, leading to disruption and cost of repair/clearance
- Poles (power, phones, etc) lean or fall, leading to disruption of services/supply
- Loss of farmland – landslides, creep of topsoil
- Damage to structures, e.g. bridges, buildings, pipes
- Quarrying and mining disasters.

Remember

Mass movement events have mainly primary impacts. The most common secondary impact is flooding, caused by a landslide blocking a river. Disruption of communications is another secondary impact. Often mass movements are secondary hazards triggered by another event, e.g. an earthquake.

Exam tips

It is possible to forecast mass movement hazards but their timing and extent can rarely be accurately predicted. Many areas have hazard maps suggesting the most likely route of mass movements. For more information about avalanche hazards, go to www.heinemann.co.uk/hotlinks, enter the express code 7719P and click on the relevant link.

Ways of reducing mass movement hazards:

- Reduce pressure on the top of the slope, e.g. limit building
- Control slope face processes, using drains, sheet piles or steel nets
- Reduce slope foot processes, e.g. gabions, revetments
- Reduce moisture content of slope, e.g. drains
- Change the angle of the slope – regrade it as a gentler slope
- Plant trees and bushes – helps bind the slope and dries it out.

Case studies: mass movements

	Aberfan, October 1966 **Student book page 12**	Swiss Alps, St Moritz, Davos, January 2007
Type	Flow/landslide – 2 million tonnes	Avalanche
Causes	◆ Waste coal tip at steep angle ◆ Prolonged heavy rain ◆ Tip built on a spring ◆ Little vegetation to bind the waste ◆ No management of the tip ◆ Assumed any slide would be slow	◆ Heavy snowfall – 75 cm ◆ Mild winter – so warmer layers ◆ Trees had been removed to help skiers ◆ Strong winds built up drifts ◆ Skiers were going off piste ◆ Reshaping of slopes for hotels
Impacts	◆ 147 killed (116 schoolchildren) ◆ 20 houses and a farm buried ◆ Huge psychological impact ◆ Loss of a generation ◆ Cost of clean-up	◆ 8 killed ◆ Some buildings damaged ◆ Damage to ski slopes ◆ High cost of rescue services ◆ Roads blocked
Responses		
Short term	Emergency rescue services	Rescue services, helicopters, etc
Long term	◆ Other tips checked and slope angles reduced ◆ 1969 Mines and Quarry Act passed to control siting of tips ◆ Over £20 million (in 2007 equivalent) in donations ◆ Very expensive clean-up (£2 million) ◆ No prosecutions	◆ Fencing off avalanche-prone areas ◆ Structures designed to slow and divert avalanches ◆ Warnings and education ◆ Setting off potential avalanches ◆ Reforestation

Quick check questions

1 Under what weather conditions do most mass movement events occur?

2 What is the slowest type of mass movement? Why does this cause few hazards?

3 What distinguishes a flow from a slide?

4 Why were so many killed in the Aberfan disaster?

5 How are avalanches set off deliberately?

6 Why are so many forests being cleared on upland slopes in developing countries when they know it leads to mass movement events?

7 In what type of area do most mass movement events occur in Britain?

1.2 What are the hazards associated with flooding?

Student book pages 16–24

Causes of river and coastal floods

Remember

Most river floods in the UK occur in August.

Physical causes of floods

- Usually by altering the balance of stores, inputs and outputs in the system
- Climate – snowmelt, heavy rain (thunderstorms), low evaporation, storm surges
- Previous weather conditions, e.g. a long period of wet weather
- Relief – very flat, low-lying area
- Drainage – density, regime, hydrograph, drainage pattern and density, channel type
- Vegetation – grass versus trees
- Rock type – permeability, porosity, water table
- Soil conditions – wet versus dry, baked impermeable by a drought
- Natural disaster – earthquake, landslide
- Subsidence – area is sinking (isostatic, removal of groundwater, etc)
- Rising sea level – global warming melting ice sheets.

Human causes of floods

- Usually by putting human activities in areas at risk and interfering with the natural cycle/hydrology
- Construction of impermeable surfaces, e.g. towns, roads
- Removal of vegetation cover – deforestation, removal of coastal marshes/mangroves, farming
- Soil erosion, leading to silting of channels and storage lakes
- Drainage – ditches, waste disposal, drains, excessive irrigation
- Changing rivers – dams, diversions, embankments, removing or adding deposits
- Constricting channels, e.g. bridges, levees.

Remember

You may have covered much of this material at AS level but at A2 the question will be asking for an evaluation rather than just a description.

Impacts of flooding

Key words

Primary impacts – immediate effects (first few days)

Secondary impacts – subsequent to primary impacts (months to years)

Tertiary impacts – long-term impacts (many years)

Primary impacts

- Deaths (humans, pets and livestock), evacuation, destruction of buildings/possessions, pollution (sewage, oil, etc), disease from polluted water, stress (loss of security).

Secondary impacts

- Transport links broken (bridges, roads, rail), cost of clean-up and replacement, loss of jobs, crops ruined (animals may starve), rehousing costs
- May take over a year to dry out buildings.

Tertiary impacts

- Flood-prone areas decline as property values decrease, difficult for owners to get insurance or sell property.

Responses to flooding

Preventative planning

- Planning – avoid flood plains and green sectors, e.g. parks
- Planned retreat – leave certain areas to flood, e.g. Somerset Levels
- Reduce surface flow, e.g. afforestation, contour ploughing.

Structural planning

- Channel modification, e.g. overflows, storage areas, dams, widen channels
- Embankments – raise them, reinforce them, put up flood gates
- Flood relief channels
- Flood barrages, e.g. on the River Thames.

Management

- Early warnings – better weather and river level forecasts, communications
- Flood insurance
- Public relief funds
- Accept the risk and live accordingly
- Ignore the risk ('head in the sand' approach).

Emergency rescue

- Rescue people and animals
- Save property and possessions
- Use sandbags, erect flood barriers, pump out property.

Remember

Look at pages 6–7 to see an alternative way of classifying responses, this time to earthquakes.

Case studies: flooding

	Coastal	River
	North Sea storm surge, 1 February 1953 **Student book pages 20–24**	Yangtze, China, June 1998 **Student book pages 16–20**
Physical causes	◆ Storm surge – strong winds from north-east pushing water south ◆ Intense depression so sea level rises (1 cm per 1 mb fall) ◆ High spring tide ◆ North Sea narrows to south ◆ Heavy rain so rivers swollen ◆ Shallow coastal waters and low-lying coast	◆ Heavy prolonged rainfall ◆ Sheer size/volume of river ◆ Rapid snowmelt in mountains ◆ Earlier monsoon ◆ Isostatic uplift in upper course
Human causes	◆ Dredging of sand offshore ◆ Neglected sea walls ◆ Global warming = sea level rise ◆ Salt-marshes reclaimed ◆ Rivers embanked ◆ Rapid urbanisation of coast ◆ Encroachment onto natural flood-holding areas	◆ 85% of watershed deforested ◆ Incomplete levees ◆ Rapid soil erosion (silts the channel) ◆ Draining of wetlands ◆ Dam construction ◆ Rapid urbanisation ◆ Encroachment onto natural flood-holding areas, e.g. lakes
Environmental impacts	Land lost to salt flooding – some abandoned	Loss of habitats
Social impacts	◆ 2000 drowned in Netherlands, 300 in UK ◆ 35 000 evacuated in UK, e.g. Canvey Island	◆ 3700 deaths ◆ 15 million lost their homes ◆ Millions fled
Economic impacts	◆ £50 million damage in UK alone ◆ Very productive crop land lost ◆ Seaside resorts hard hit, e.g. Southend ◆ Coastal routes cut	◆ US$26 million loss ◆ 25 million ha of cropland damaged ◆ Industry damaged ◆ Loss of navigation on river
	Longer-term, as salt water and unexpected	Short-term impact, as used to river floods

Case studies: flooding *(continued)*

	Coastal	River
Responses	Democratic	Centralised
Short term	◆ Evacuation of people and livestock ◆ Search and rescue ◆ Property pumped out ◆ Army called in to seal sea walls	◆ Evacuation of people and livestock ◆ Search and rescue ◆ Army called in to seal levees
Long term	◆ Early warning system with sea level monitoring ◆ Sea walls raised or managed retreat accepted ◆ Areas allowed to flood to take pressure off walls ◆ Flood barrages on rivers, e.g. the Thames	◆ Increased water storage, e.g. Three Gorges Dam ◆ Reforestation – five-year programme in upper catchment ◆ Stricter control of felling ◆ Repair and strengthening of levees ◆ Better warning systems ◆ Relocation of villages at risk

1.3 What are the hazards associated with earthquakes and volcanic activity?

Student book pages 24–33

Tectonic processes

Student book pages 24–25

The tectonic plates that make up the Earth's crust move about on convection currents in the mantle. At their edges they may:

Converge

- Destructive margin (collision) = deep earthquakes, explosive eruptions
- Destructive margin (subduction) = deep earthquakes, trench

Diverge

- Constructive margin = shallow quakes, lava flows and shield volcanoes, rift valleys

Slide past

- Conservative margin = shallow quakes, hot springs, few eruptions

Key words

Subduction
Plate tectonics
Plate boundary

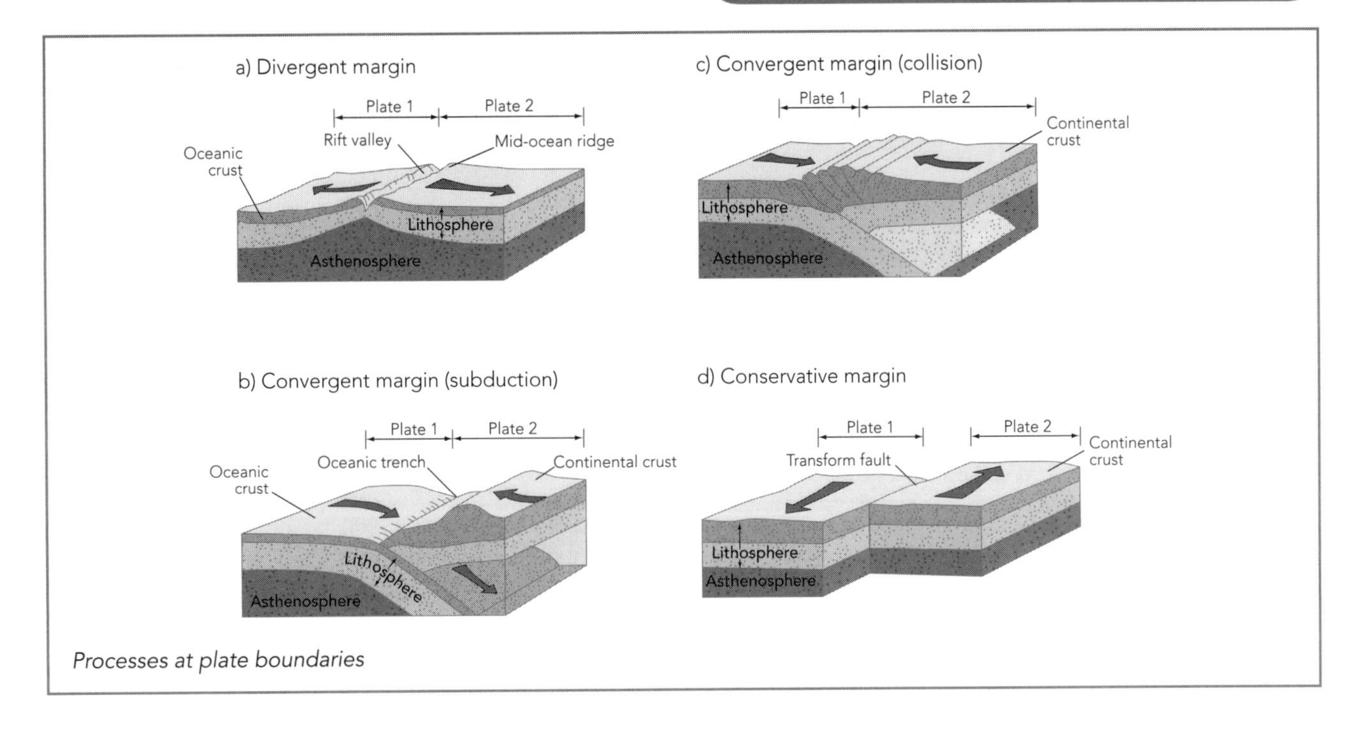

Processes at plate boundaries

Remember

It is very unlikely that you will be asked directly about plate tectonics, but tectonics does help to explain why margins differ in their characteristics and range of potential hazards. The lighter, less dense, but thicker continental crust tends to be 'pushed around' by the denser, thinner oceanic crust.

Hazards and impacts of earthquakes

Student book pages 24–9

Hazards have the **potential** to cause loss of life and/or damage to property (so people need to be present).

- Primary hazards – ground shaking, landslides, faulting at surface
- Secondary hazards – liquefaction, ground failure, rock falls, mud flows, tsunamis.

Impacts are what actually happens. There is a contrast between the immediate disaster and the long-term impacts; the scale of the impact is crucial.

- Primary impacts – destruction, casualties, landslides, fires, loss of services/utilities and communications, shock and traumatic stress, violence (looting)
- Secondary impacts – disease (sewage and water pipes broken), loss of infrastructure, housing, jobs, food and water shortages, floods
- Tertiary impacts – cost of recovery, loss of crops, damage to mines/industries, trade, long-term depression, out-migration.

Responses

- Emergency aid – rescue, tents, medicines, food, water, often involving the military and charities
- Short-term aid – rehousing people, rebuilding hospitals, repairing infrastructure
- Long-term aid – moving population, improving warning systems, emergency planning.

Key words

Focus

Epicentre

P and S waves

Case studies: earthquakes

	Earthquake in a LEDC	Earthquake in a MEDC
	Kashmir, October 2005 **Student book page 29**	Los Angeles, Northridge, 1994 **Student book page 26**
Magnitude	7.7 and 900 aftershocks	6.7, lasting 15 seconds, and many aftershocks
Time	9.20 am – schools full	4.30 am – so few people about
Physical causes	◆ Australian plate colliding with Eurasian plate ◆ Mountainous area ◆ Remote ◆ Extreme climate – winter snows	◆ A previously unknown buried thrust fault ◆ Steep hillsides
Human causes	◆ Poverty ◆ Poor-quality buildings ◆ Disputed area between India and Pakistan	◆ Building on the fault line ◆ Many older buildings ◆ Bridges failed so difficult to get aid in
Environmental impacts	◆ Landslides ◆ Diseases – tetanus, cholera ◆ Contaminated water ◆ Cold killed many	◆ Landslides ◆ Fires ◆ Disease – respiratory from spores (valley fever)
Social impacts	◆ 73 000 killed and 69 000 injured ◆ 3 million left homeless ◆ Shock and stress ◆ Young generation wiped out	◆ 66 killed and 9000 injured ◆ Homes damaged ◆ Looting in poorer areas ◆ Shock and stress

Case studies: earthquakes *(continued)*

Economic impacts	◆ US$5 billion of damage ◆ 1442 schools collapsed ◆ Bridges and roads destroyed so some areas left isolated for weeks ◆ Wells destroyed ◆ Food stores destroyed ◆ Phone and power lines broken	◆ US$30 billion of damage ◆ 9 bridges collapsed ◆ 11 major roads closed ◆ 11 hospitals closed ◆ Major car parks collapsed ◆ Water, power and sewer pipes broken
Responses	Slow, as remote	Fast, as prepared
Short term	◆ Search and rescue – slow reaction by army ◆ Lack of tents, food, etc	◆ Search and rescue ◆ Emergency centres opened ◆ National Guard called in
Long term	◆ Large international aid effort (but less than for tsunamis, as little-known area) ◆ Political unrest at failure of government to act and its corruption ◆ Rebuilding has been slow ◆ Still military tension in area	◆ Reconstruction ◆ Laws – by January 2005 all hospitals had to have earthquake-proof A and E rooms ◆ California state set up its own low-cost earthquake insurance ◆ Bridges reinforced ◆ Vulnerable buildings inspected and reinforced

Hazards and impacts of volcanoes

Student book pages 30–33

Hazards

- Primary – landslides, fire, mud flows, rock bombs, lava flows, gas, heat, ash, pyroclastic flows
- Secondary – tsunamis, crop failure, disease, famine.

Impacts

There is a contrast between the immediate disaster and the long-term impacts; the scale of the impact is crucial.

- Primary – destruction, casualties, landslides, fires
- Secondary – disease, loss of infrastructure, housing and jobs, food and water shortages
- Tertiary – cost of recovery, loss of crops, damage to mines/industries, trade.

Responses

- Emergency aid – rescue, tents, medicines, food, water, often involving the military and charities
- Short-term aid – rehousing people, rebuilding hospitals, repairing infrastructure
- Long-term aid – moving population, improving warning systems, emergency planning.

Key words

Lahars	Pyroclastic flows
Tephra	Tsunamis

Case studies: volcanic eruptions

	Eruption in a LEDC	Eruption in a MEDC
	Montserrat, Caribbean island, 1995–97 **Student book pages 31–33**	Mount St Helens, USA, May 1980
Basic cause	Atlantic plate colliding with Caribbean plate	Pacific plate colliding with North American plate
Physical causes	◆ Stratovolcano with lava domes ◆ Pyroclastic flows (1997) ◆ Ash clouds – 40 000 ft in height 600 000 tonnes (1996) ◆ Landslides ◆ Domes collapse (1995)	◆ Level 5 earthquake ◆ Landslide = lateral blast ◆ Pyroclastic flow ◆ Vast ash cloud (20 km high) ◆ Snow melt = lahars (145 km/h)
Human causes	◆ Developing country with little technology ◆ Poor country	◆ National Park so largely ignored ◆ Wildness meant few roads

Case studies: volcanic eruptions *(continued)*

	Eruption in a LEDC	Eruption in a MEDC
Human causes (*continued*)	◆ Relatively high population density ◆ Bulk of communications radiated from main volcanic area ◆ Capital and airport too near volcano	◆ Very low population so often difficult to find ◆ Bridges vulnerable to flash floods
Impacts	Long build-up, with earthquakes in 1992	◆ 2 months warning build-up and very closely monitored ◆ Knew past historical pattern of eruptions
Environmental impacts	◆ Volcanic dust caused silicosis and asthma ◆ Coral reefs buried in ash/dust ◆ Cloud forest destroyed ◆ Acute acid rain ◆ Lakes with pH 1.5 – death of aquatic life	◆ Fish killed in hot, choked rivers ◆ Wildlife wiped out, e.g. 5000 deer ◆ Thousands of trees blasted over 600 km^2 ◆ 540m tonnes of ash fell on 60 000 km^2
Social impacts	◆ 19 deaths (all farmers) ◆ Southern 60% of island left uninhabitable – loss of homes ◆ 50% of population went to UK or USA	◆ 57 killed ◆ 200 homes lost ◆ Little stress/shock recorded
Economic impacts	◆ Airport destroyed ◆ Capital (Plymouth) buried under 12 m deep mudflow ◆ Loss of main roads, homes and other infrastructure ◆ Loss of main flat fertile farm area with its cash crops ◆ Loss of once large tourist trade	◆ US$3 billion damage (2007 prices) ◆ Ash made roads slippery ◆ Lahars blocked roads ◆ Crops destroyed, as ash choked pores in leaves of plants ◆ Closed navigation on Columbia river ◆ 47 bridges destroyed ◆ 300 km of roads blocked ◆ Visibility poor for 2 weeks, forcing airports to shut ◆ Ash caused power cuts and blackouts
Responses	Lacked local resources – needed help from UK	Effective, as hazard was known and expected
Short term	◆ Evacuation to safer north ◆ Exclusion zones set up (60% of island)	Search and rescue but area already evacuated
Long term	◆ 1998: UK granted residency rights in UK to islanders ◆ UK launched three-year $122 million reconstruction programme ◆ EU gave $12 million grant for relocating capital to Brades estate ◆ New airport opened in 2005 ◆ Increased monitoring of volcano ◆ Decline of once prosperous island ◆ House building and diversification of economy into services (80%)	◆ Clean-up cost $363 million ◆ Increased tourism, as now nationally known ◆ Increased employment during clean-up ◆ US forest service took over area as a National Volcanic Monument ◆ Long-term monitoring and research

Quick check questions

1 Which type of tectonic margin represents the greatest hazard?
2 Why is fire so likely after an earthquake?
3 Why are aftershocks, following an earthquake, so dangerous?
4 Why did the Kashmir earthquake victims get comparatively little help?
5 Why are the military used in such natural disasters?
6 Why are volcanic hazards more easily planned for than earthquakes?
7 Why was the death toll so low in both the Montserrat and Mount St Helens eruptions?
8 Why do volcanic eruptions affect areas way beyond the immediate site of the eruption?

1.4 Why do the impacts on human activity of such hazards vary over time and location?

Student book pages 34–37

Remember

Human activity includes SPITMAS (settlement, power, industry, transport, mining, agriculture, services).

Impacts on human activities vary over **time** according to:

- Level of development/technology
- Recurrence interval and frequency of event
- Time of day, season, year
- Build-up of events – warning interval.

Impacts on human activities vary with **location** according to:

- Distance from plate margin, river, slope, etc
- Ability to predict or forecast (level of technology)
- Population density, distribution, level of perception and education
- Level of communications – warning systems
- Mobility of population
- Highland versus lowland
- Urban versus rural
- Level of development – building type, ability to warn/evacuate
- Remoteness
- Type and size (magnitude) of hazard or mix of hazards
- Geology and rock structure – weak rock is more vulnerable.

Exam tips

In an exam you may be asked a general question about earth hazards. Always try to compare at least three types of hazards to bring out the differences in scale, predictability, impact, etc.

Differences between floods, earthquakes and volcanic eruptions

	Floods	Earthquakes	Volcanic eruptions
Possible to predict?	Quite possible, e.g. by checking rainfall level	Virtually impossible (seismic gap idea)	Fairly good if monitoring equipment in place
Known hazard?	Usually a history of flooding	Not always and not exactly located	Usually obvious, cone-shaped mountain, etc
Warning?	Usually some warning – up to a few days	Often no warning	Usually precursor events, e.g. swelling, earthquakes, etc
Scale?	Vary from local to regional	Regional	Local to regional
Duration?	A few days	Seconds	Days to years
Impacts?	Variable – depends on size of river	Severe but depends on strength and depth of quake	Variable – depends on scale of eruption and type of volcanic material
Area of impacts	From local to regional	From local to regional	Local to global (ash cloud)
Short term	◆ Deaths ◆ Destruction ◆ Disruption	◆ Deaths ◆ Destruction ◆ Disruption	◆ Few, as can usually evacuate people ◆ Destruction
Long term	◆ Few long-term problems, as floods recede and can rebuild ◆ Long-term gain as floods leave fertile soil	Lot of medium-term problems (reconstruction) but few long-term problems	Problems for years as landscape is changed, but eventually nature and economy recover
Relative impact in deaths per million from 1990 to 1999	Asia: 14 Africa: 10 N. America: 1	Asia: 23 Africa: 2 N. America: 1	Asia: 0.00005 Africa: 0.000003 N. America: 0.000002

Is the impact always greater in a LEDC?

Here is a comparison of a volcano in a rural developing country (the Philippines) versus an earthquake in an urban MEDC (Japan).

Volcano: Mount Pinatubo, Philippines, June 1991

Over 100 000 people lived on the slopes of the dormant Mount Pinatubo in the Philippines. However, when the volcano erupted only 350 were killed by collapsing buildings (and few died directly because of the eruption). Even though this was the second-largest volcanic event in the 20th century, which sent lahars, ash and rock over a 100 km radius, there was a low death rate for a developing country because:

- The earthquakes started in mid-March, giving an early warning
- The US Geological Survey started accurate monitoring of the volcano and the US Air Force took gas samples at altitude
- Tiltmeters indicated that the mountain was inflating in early June
- A hazard map was drawn up and evacuation zones established
- Five stages of alert were established
- There was good communication, both directly with locals and via the media
- The US Army helped with the evacuation (using two local US bases).

Unfortunately, the eruption coincided with a tropical storm, which led to lahars and increased the weight of the ash. More people died from disease in the overcrowded refugee camps than were killed by the eruption.

Earthquake: Kobe, Japan, January 1995

This was a 7.3 level quake (compare with Northridge, pages 6–7), which killed 6434 and did US$200 billion of damage, or 2.5 per cent of Japan's Gross Domestic Product (GDP). Casualties were greater than expected for a MEDC because:

- The earthquake struck at 5.46 am
- Early warning systems failed to pick it up
- The quake was close to the surface and near the city
- The rock type magnified the shock and some caused liquefaction
- There were few reinforced buildings
- Traditional wooden buildings caught fire (more people died in the fires than the quake)
- Expressways collapsed, hampering rescue attempts
- There was a lack of co-ordination of rescue services
- The Japanese government initially refused international help
- Kobe was a large urban area with a population of 2 million.

Exam tips

It is too simplistic always to assume that a country's level of development is the key to explaining differences in the level of hazard impact. Usually earthquakes will have larger impacts than volcanoes, wherever they are, because earthquakes are so unpredictable.

1.5 How can hazards be managed to reduce their impacts?

Student book pages 37–41

Stages in reducing impacts – management strategies used

1. **Prediction** – based on past events (hazard mapping), monitoring of gas/pressure etc, tiltmeters, chemical sensors, ultrasound
2. **Risk assessment** – calculate size and extent of risk, inform population, building design, location of vital buildings/facilities, e.g. power stations
3. **Prior prevention** – lubricate fault lines, afforest slopes, raise and strengthen levees, move vulnerable population and livestock, establish exclusion zones
4. **Planning** – individual (e.g. store water), local authority (e.g. set up emergency centres), state or central (e.g. mobilisation of rescue services, building controls)
5. **Preparation** – education (emergency drills), contingency plans (e.g. evacuation routes signposted, training of emergency staff)
6. **Warnings** – use of media, communications, level of threat, planned evacuations
7. **During prevention** – dam or divert lava flows, flood volcanic vents, stabilise slopes, divert rivers

8. **Response** – search and rescue, emergency aid, insurance, state or international aid for rescue, relocation
9. **Recovery** – clearance of debris, state aid for reconstruction, tax relief
10. **Redevelopment** – long-term plans.

Exam tips

The contrast in the Pinatubo and Kobe impacts illustrates the relative effectiveness of management strategies.

Adjustments that can be made to reduce hazard impacts

Physical adjustments

- Altering characteristics of hazards e.g. in Iceland (Heimaey lava flow halted by spraying with seawater)
- Building to withstand hazards, e.g. earthquake-proof structures in California
- Constructing diversions, barriers, etc, e.g. on Mount Etna in 2001 to divert lava
- Moving people to less vulnerable locations, e.g. evacuating them from Mount Pinatubo.

Social adjustments

- Public awareness via education, media, etc, e.g. Japan's annual earthquake day
- Evacuation plans and preparations, e.g. California's household regulations, in case of an earthquake
- Greater community involvement to reduce vulnerability, e.g. following Kashmir earthquake.

Political adjustments

- Evacuation plans, services and emergency centres, e.g. these were increased in the Indian Ocean area following the 2004 tsunami
- Land use zoning and restrictions, e.g. along the River Avon in the West Midlands (UK) to reduce flood risk
- Issuing early warnings and co-ordinating emergency services, e.g. in the 2007 Tewkesbury floods (UK)
- Spreading economic loss through insurance, grants, etc, e.g. UK government following Montserrat eruption.

Remember

The public's perception or awareness of potential hazards is crucial. People will often offset the threat by considering the benefits of living in such areas, or they may not be able to afford to move elsewhere.

Managing earth hazards

Managing earth hazards effectively depends upon:

- The nature of the hazard – its severity, scale, frequency, any build-up signs, etc
- The level of preparation and awareness of risks
- The nature of the area – its structure, geology, relief, climate, etc
- The level of development – research, technology available, communications, warnings, etc
- The nature of the population – density, education, mobility, level of perception, etc
- Political organisation – co-ordination, priority, existence of emergency plans, etc.

How was the 1973 eruption in Heimaey, Iceland, managed?

Background

In January 1973 a 2 km-long fissure eruption took place on the island of Heimaey, Iceland (a MEDC). Lava flows threatened the main town and the port.

Management successes

- All 5300 residents were evacuated (80 per cent returned by 1975)
- Only one man died (from fumes)
- Trenches and walls were built to funnel lava and gases
- Seawater was used to cool lava flows (43 pumps were used and 6 million cubic metres of water)
- The port was saved (25 per cent of Iceland's fishing fleet)
- Iceland now uses the heat to generate geothermal power (40 megawatts)
- Lava and ash have added to the land, creating a larger area on which to build 200 new homes.

Reasons for management success

- Eruption lasted from January to April – there was time to react
- Viscous lava is so slow-moving
- Hazard and risk were known
- Ease of evacuation by sea and air

- Small, educated population willing to evacuate quickly
- MEDC so they had technology to cool the lava
- Wealthy country – it cost US$1.5 m to cool lava
- Politically sophisticated so management well co-ordinated.

Exam tips

Questions may ask you to evaluate the relative success of a range of strategies used to manage earth hazard events. You must try and explain why some strategies are more successful than others. This will partly depend on the nature of the event and the location in which it took place.

Quick check questions

1. Why is a hazard event at night more dangerous?
2. Why is an urban area more vulnerable than a rural area in an earthquake?
3. What is the difference between solid bedrock and sediments when hit by an earthquake?
4. Why do you think Africa suffers more deaths each year from floods than from earthquakes?
5. How do tiltmeters help detect a forthcoming volcanic eruption?
6. What is usually used to lubricate a fault line? Does it work?
7. How was the lava flow halted during the Heimaey eruption?
8. What has been done to reduce the likelihood of another Indian Ocean tsunami disaster?

ExamCafé

Student book pages 6–45

Section A

Sample question

Study resource 1: a diagram of a railway cutting in a tropical LEDC, showing a mix of sands and clays with a little vegetation and a steep slope that is cut into by the railway line.

Identify ***one*** *issue and suggest appropriate management strategies to deal with it.*
[10 marks]

- This is an earth hazards question so the cross-section implies there is a possible hazard here. Clearly, it would be a mass movement one.

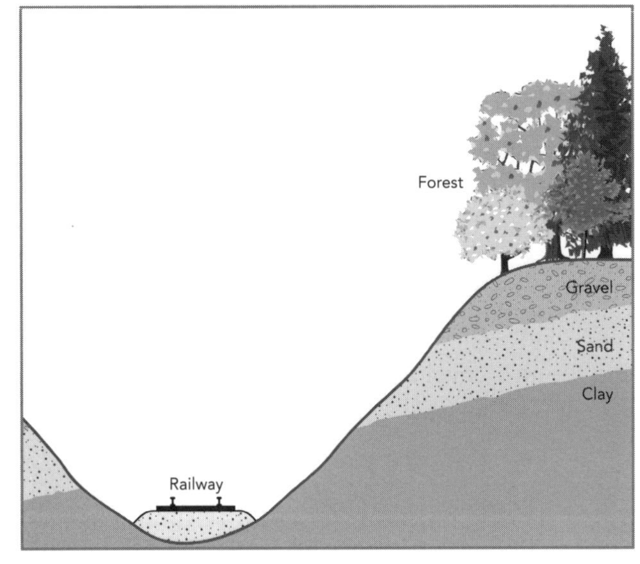

Resource 1: Cross-section diagram of railway cutting in a tropical LEDC

Student Answer

The cross-section shows the classic situation for a landslide or mudflow to occur to block the railway.

This is fine but has it fully identified the nature of the issue? Another possible answer would be:

Tropical areas get heavy rain and this will add weight and lubricate the impermeable clay areas. As the slope is steep and lacks any vegetation to hold it in place it will slip easily, possibly triggered by the vibrations of the train.

This is much more detailed and shows clear understanding of the conditions for mass movement. This student also uses effective and correct geographical terminology. As for appropriate management strategies, there are a number of possibilities but it is a LEDC so funds will be limited. (This illustrates the importance of understanding the significance of key words in the question.)

So the student might go on to say:

A strategy would be to pile-drive interlocking steel plates into the slope. These would then hold the slope in place and drains could be inserted every so often to take off surplus water.

Or alternatively:

Labour is probably cheap and plentiful so the slope angle could be reduced to nearer its stability angle and then planted with fast-growing shrubs to hold the slope in place. Their transpiration would reduce the water content.

Both are valid strategies but the latter is more appropriate for the scenario of a LEDC so would gain more credit.

Section B

Sample question

'The severity of the impact of an earth hazard on an area reflects the area's level of development rather than the nature of the hazard event.' To what extent do you agree with this statement? [30 marks]

Like all evaluation questions there is some truth in this but equally the statement can be contested. You should always start by reading the question carefully. There are several key words here, each of which raises questions:

- **Severity:** in what terms – deaths, damage, etc?
- **Impact:** primary, secondary, tertiary, or short term versus long term, or environmental, economic or social?
- **Area:** what size or scale?
- **Level of development:** economic or socio-political?
- **Nature:** size, duration, type, severity?
- **Hazard event:** mass movement, flood, earthquake, tsunami, volcanic eruption?

Suddenly the question seems much bigger. Can you provide a succinct answer in the 45 minutes you probably have for this question? If not, don't attempt it. This is a good example of where an essay plan would help you sort out a possible approach:

Student Answer

Sample plan

Introduction – define terms

An example where it is true, e.g. Kashmir versus Northridge earthquakes

An example where it is not true, e.g. Kobe earthquake – scale of hazard

Hazards differ in impact in same country, e.g. Aberfan versus North Sea floods

Impacts differ as hazards differ, e.g. earthquake – no warning/volcano – warning

Severity differs, e.g. physical versus economic versus social

This plan looks promising, albeit a tight fit in the time. So what would the conclusion be?

Conclusion

Broadly speaking the statement is true, as the level of development of an area does clearly influence the level of technology and resources available to predict, prepare for and warn of hazardous events. However, extreme events will cause severe impacts regardless of the level of development. Kobe demonstrated that a highly developed, highly prepared area where earthquakes are expected can still be devastated. Equally, it may be the nature of the area rather than its level of development that makes the severity of the impact worse – would the Indian Ocean tsunami have killed so many if the coastlines had been higher?

This is an effective conclusion – it revisits the question, answers it and offers two qualifications of the evaluation.

Chapter 2 Ecosystems and environments under threat

Student book pages 46–83

2.1 What are the main components of ecosystems and environments and how do they change over time?

Student book pages 48–55

Remember

If you have chosen this topic you may wish to revisit some of your AS physical unit's case studies of environments.

Main components of ecosystems and environments

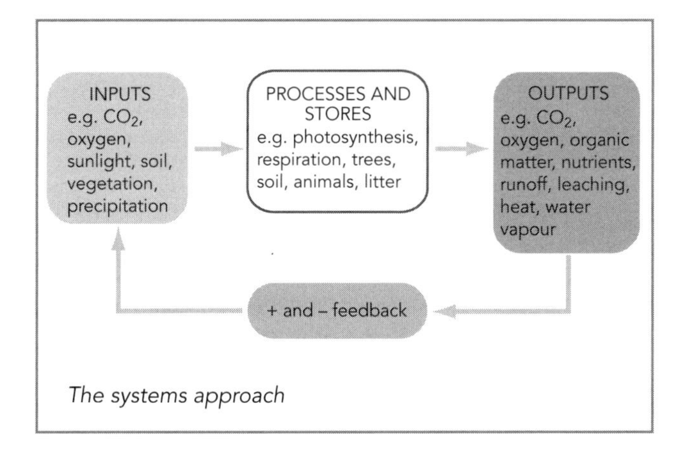

The systems approach

- **Stores** – biomass (plants, animals, etc), litter and humus, soil
- **Flows** – leaf fall, death, decomposition, nutrients drawn up
- **Inputs** – water, sunlight, minerals from weathering, litter/minerals washed/blown in, deposition from elsewhere
- **Outputs** – leaching from litter and soil, erosion, evapo-transpiration

If one store or flow changes, so do all the others. This results in a change in the characteristics of the ecosystem or environment, which is dynamic (ever changing).

Key words

Gross primary productivity (GPP)
Net primary productivity (NPP)
Photosynthesis
Biomass

Trophic levels

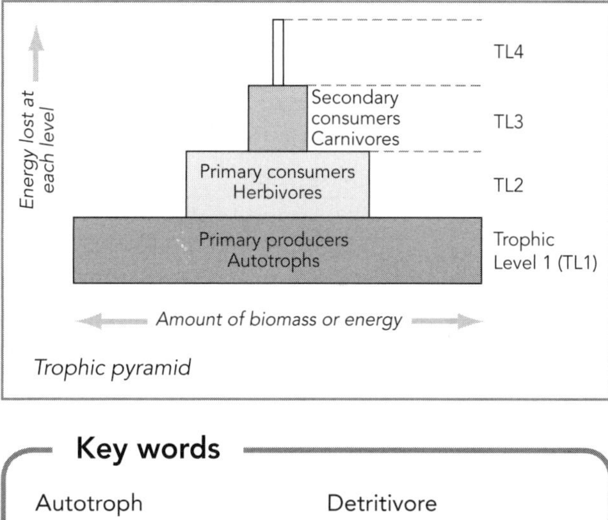

Trophic pyramid

Key words

Autotroph
Omnivore
Trophic level
Detritivore
Saprophyte

Moving up the trophic levels, energy (nutrients) or biomass is lost because:

- Not all food can be digested (e.g. bones and lignin)
- Not all plants/animals are eaten
- Activity (e.g. eating) consumes energy
- Heat energy is lost through respiration and excretion
- Body mass radiates heat – therefore energy is lost.

This means that there is a reduction in the total biomass and number/variety of species as we move up the trophic levels.

Exam tips

A typical issues question in section A would ask you to identify an unbalanced trophic pyramid.

Remember

Remove one organism from a chain and there will be a knock-on effect – both up and down the chain.

Nutrient cycles

The diagram opposite shows how an individual ecosystem transfers its nutrients – the relative size of stores and flows is crucial.

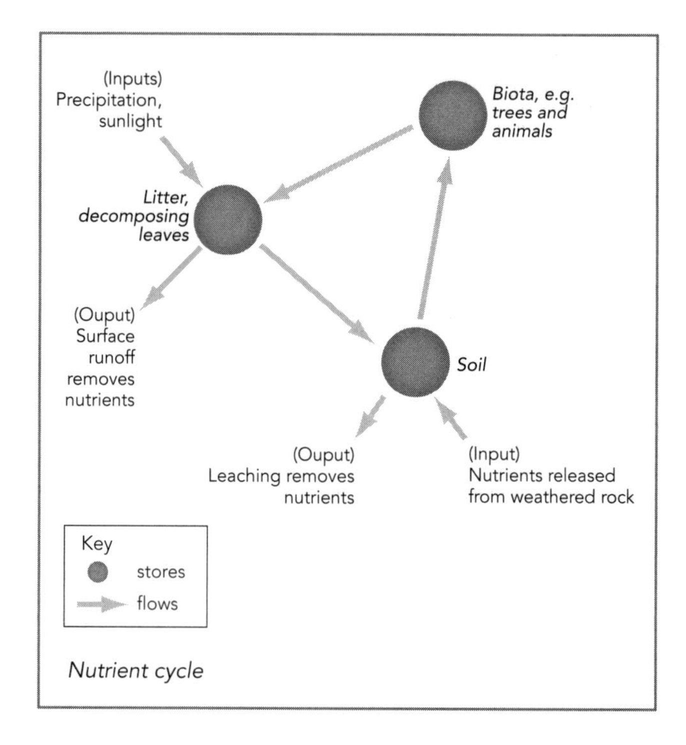

Nutrient cycle

Case studies: woodland ecosystems

	Deciduous woodland	Coniferous woodland
Location	Middle latitudes (30–40° N)	High latitudes (40° N)
Net primary productivity (NPP)	High (1200g/m²/yr), as high summer temperature	Low (800g/m²/yr), as cold winters and dark
Inputs	High precipitation, sunlight, weathering	Low precipitation, summer sun, slow weathering
Outputs	Heavy leaching	Heavy leaching (snowmelt)
Stores		
Biomass	High (35 kg/m²)	Quite high (20 kg/m²)
Litter	Low, as deciduous and easily decomposed	High, as evergreen, acid tough needles so slow breakdown
Soil	High, as quick decomposition and lot of soil organisms	Low, as slow decomposition and few soil organisms
Flows		
Biomass to litter	High in autumn, as deciduous	Low, as evergreen
Litter to soil	High, as decomposes easily	Low, as slow to decompose
Soil to biomass	High, as active, deep roots	Low, as shallow roots active in summer

Nutrient stores and flows reflect the climate, which supplies inputs of water and energy:

	NPP kg/m²/yr	Biomass kg/m²
Tundra	140	0.6
Mediterranean scrub	700	35
Tropical rainforest	22 200	45

Remember

This is a synoptic paper so you can and should refer to the cold or hot arid ecosystems you studied at AS level.

Change in ecosystems and environments

How does change occur?

Natural succession (without human interference)

From bare rock/soil, through a number of stages (or seres), to climax community, which is in balance with the climate. Stages are controlled by micro-climate, nutrient supply, soil acidity and competition.

Generic or typical succession

SERES	1	2	3	4	CLIMAX
Biomass	Low	Increasing	Increasing	Increasing	Static
Variety of species	Few – tough pioneers	Increases colonisers, rapid growth	Increases secondary colonisers, competition	Slows, dominants emerge larger	Decreases, dominants out-compete
Nutrients available	Low	Increasing	Increasing	Decreasing	Decreasing

But humans can interfere and hold an ecosystem at an artificial climax (plagioclimax) e.g. by mowing grass.

Or a natural disaster such as a flood can produce a **sub-climax**, which then undergoes a secondary succession to reach climax.

Key words

Sere
Climax
Sub-climax
Succession
Plagioclimax
Pioneers

What environmental factors change?

- Inputs changed, e.g. climate
- Outputs increased or decreased
- Nutrient stores
- Flows – speeded up, slowed or disrupted
- Internal structure, e.g. trophic levels changed.

How do physical and human factors affect environments?

Physical factors are less likely to cause change, especially sudden change. This is because ecosystems can usually adapt over time, as physical factors rarely change suddenly. (Exceptions would include floods, natural disasters, etc.)

Physical factors	Effects
Climate	Changes inputs (e.g. precipitation) and outputs (e.g. evaporation). Also speeds up flows (warmth) or slows them down (cold)
Drainage	Lowers water table, dries soil, *or* floods, waterlogs and drowns species
Natural disaster	Can cause dramatic change (e.g. eruption or hurricane)
Fire	Wipes out systems but some species need fire to germinate
Human factors	**Effects**
Settlement	Clears vegetation and creates impervious surfaces
Trampling	Can crush plants, also compacts the soil and reduces its porosity
Agriculture	Adds chemicals, introduces new species or removes them (e.g. weeds), compresses the soil, grazing, drains wetlands or irrigates
Transport	Pollution – changes soil acidity, kills vegetation
Mining/Quarrying	Removal of surface vegetation and soil, exposes fresh surfaces
Power production	Burns fossil fuels = acid rain, global warming, etc
Drainage	Lowers water table, dries soil

Quick check questions

1. Which flow in the nutrient cycle links the soil to the biomass?
2. Why are different trophic levels never the same size in biomass?
3. Why do deciduous forests produce more biomass than coniferous forests?
4. How much greater is the biomass per square kilometre in a tropical rainforest, compared to tundra?
5. How does nature supply bare rock surfaces to initiate a succession?
6. Why do human factors have a greater impact on succession than natural physical factors?

2.2 What factors give the chosen ecosystem or environment its unique characteristics?

Student book pages 55–62

Exam tips

The examiner wants to see detailed information that shows you have studied a local area. A locational map is a useful way to do this.

Remember

Quote some real figures for climate and try to find out about the local soils and underlying geology.

Possible factors include:

Physical

- Micro or local climate – precipitation, temperature, humidity, wind, sunshine hours
- Relief – slope angle/direction, altitude, roughness
- Rock type – porosity, minerals, structure, pH
- Drainage – free draining versus waterlogged
- Soil – depth, structure, texture, minerals, pH
- Biotic – plants, animals, etc.

Human

- Agriculture – chemicals, adding species, drainage, irrigation, clearance
- Settlement – micro-climate, drainage
- Industry – and transport
- Pollution
- Conservation and recreation.

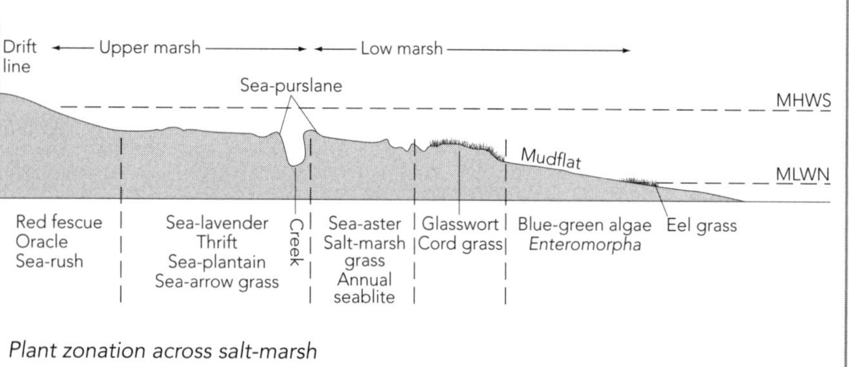

Plant zonation across salt-marsh

Key words

Exotic species
Gleys
Microclimate
Halophyte
SACs (Special Areas of Conservation)
SPAs (Special Protection Areas)
SSSIs (Sites of Special Scientific Interest)

Case study: the Solent – a salt marsh

Student book page 61

Characteristics	
Location	Salt-marsh develops between low and high water levels in sheltered shallow estuaries. It floods and drains with the tide.
Plants	Salt-tolerant plants (e.g. glasswort and cord grass) trap silt.
Wildlife	Dominated by birds, especially waders (e.g. redshank and geese). Few land animals but many fish and crustaceans.
Physical factors	
Microclimate	High rainfall washes out salt; low rainfall and high temperatures increase it. Must be able to withstand flood and drought.
Soil	Gleys (waterlogged clays and silts) so many plants have shallow roots.
Relief	Very low and flat so poor drainage and creeks and pools form salt-pans. Ever-shifting mud so some plants (e.g. cord grass) have deep roots.
Drainage	Twice-daily tides and seasonal differences, so salt resistant (**halophytes**). Animals seal their burrows or regulate body fluids.
Succession	Result of salt-resistant pioneers binding the mud and adding nutrients when they die. Organic matter gradually raises height, so drier, less resistant species colonise.

Case study: the Solent – a salt marsh *(continued)*

Human influences	
Industry	Marshes reclaimed for industrial sites (e.g. Fawley refinery and power station) and outfall of hot water has raised temperature in Solent, creating an environment for exotic species. Marshes also used for cockles and oysters.
Transport	Port development – container terminal at Southampton, oil jetties at Fawley = increased dredging and ship wash. Road construction (e.g. M275) on reclaimed marsh.
Pollution	Industrial and residential effluent (e.g. oil spills).
Agriculture	90 per cent of land is drained for grazing – light grazing diversifies but heavy tramples and destroys. Runoff from nearby fields often has an impact = algae growth.
Settlement	Urban expansion of Portsmouth, Southampton and Chichester – increases runoff and pollution (sewage).
Recreation	Creation of marinas (over 10 in area, e.g. Hamble river) – wash from boats damages marsh and anti-fouling paint gets into food chain.
Conservation	4 SACs, 4 SPAs, National Nature Reserve, 5 Local Nature Reserves, many SSSIs and bird reserves owned by RSPB (e.g. Dibden Bay).
Other	Salt-marshes act as an important coastal defence – a buffer against storm waves and a defence against the threat of rising sea levels due to global warming.

2.3 In what ways are physical environments under threat from human activity?

Student book pages 62–66

Types of threats to environments

- **Extinction** – destroying species directly or indirectly. The natural rate seems to be 10–25 species a year but, of the 40 177 current species, 784 are extinct and 16 119 are listed as threatened. Every 20 minutes another species becomes extinct. Half of all species may be extinct by 2100.
- **Coextinction** – when one species becomes extinct, others reliant on it (e.g. parasites) become extinct as well.
- **Modification** – changing the balance of flows in the system or adding or removing species in the system.
- **Alteration** – increasing or decreasing elements in the system.
- **Preservation** – keeping a system static and not allowing natural change.

Impacts of planned and unplanned human activities

Planned activities

- Agriculture – woodland clearance, field drainage, fertilisers (nutrient enrichment), pesticides and herbicides (remove species), removal of hedgerows (nest sites)
- Mining and quarrying – strip (open-cast) mining/quarrying removes surface vegetation (e.g. new planned coal mine in Yorkshire), waste heaps create new environments: steeper, faster-draining, different minerals (e.g. White mountains of Cornwall and waste from china clay pits – mostly sand)
- Construction – removal of vegetation for building homes, industry, etc; even if not cleared the ecosystem is disrupted by noise, drainage, etc (one of the arguments against planned new eco-towns in the UK)
- Forestry – by its nature this destroys trees but also undergrowth; selective felling is better than strip felling (e.g. Amazon rainforest) where everything is bulldozed
- Transport – road building or widening (creates barriers to movements and pollution from traffic fumes, also alters drainage), airport building, dredging for ports.

Unplanned activities

- Introductions – new species (e.g. Japanese knotweed) that out-compete with native ones, or alter conditions (e.g. conifers make soil more acid and reduce light reaching woodland floor)
- Fire – increased fires due to vehicles, tourists, etc – often where vegetation isn't adapted to fires, e.g. deciduous forests such as the New Forest in Hampshire
- Disease – Dutch Elm Disease brought in from Canada on imported logs in 1970s reduced species but also had a knock-on effect, as it changed the woodland canopy
- Disruption of food chain – human activity often interrupts nutrient flows (e.g. removal of wolves in the Rockies removed major carnivore so grazers increased, damaging vegetation)
- Floods – often a result of building on flood plains so increasing runoff (e.g. increased frequency of floods along River Avon has reduced burrowing species but increased growth of willows)
- Removals – ending of long-term farming practices (e.g. grazing of pigs in forests) so controls removed
- Pollution – acid rain (kills vegetation and reduces soil micro-organisms), global warming, smog (reduces photosynthesis), noise from recreation (e.g. motorbike scrambling).

Key words

Extinction | Coextinction
Modification | Preservation

Remember

Much human interference in ecosystems is unplanned and reflects our lack of knowledge of the 'knock-on effects' of one action on other elements or areas of the system. (Think back to your AS examples.)

Case study: Epping Forest

Student book pages 65–66

Woodland	
Area	◆ 2400 ha (18 km long by 4 km wide) on eastern side of London between River Lea and River Roding ◆ Largest open space in London
Ecosystem	◆ 70% deciduous woodland – beech, oak and hornbeam with understorey of holly and yew ◆ Also heath 4%, bogs and ponds 6%, grassland 20% ◆ Home to fallow deer and muntjac and rare invertebrates (e.g. newts) ◆ Has largest stock of ancient trees (over 400 years old) in UK
History	◆ Norman hunting ground and commoners grazed livestock ◆ 1878 – Epping Forest Act gave ownership to Corporation of London ◆ Since the 1981 Countryside Act, 66% is now SSI or SAC
Threats	
Natural	◆ Former open grazing now being re-colonised by secondary woodland ◆ Once pollarded trees are increasing in canopy, so shading out species ◆ Infilling of ponds by vegetation ◆ Deer damage trees (bark stripping) ◆ Global warming – increases pests ◆ Fluctuating water levels (may reflect increased runoff)
Human	◆ Recreation – this is one of the 'green lungs' of London and is very accessible and used for walking, horse riding and mountain biking ◆ Pressure for car parking spaces ◆ Lack of management until recently ◆ Vandalism (e.g. fires) ◆ Pollution – especially litter but also acid rain; car exhausts have reduced lichen species from 150 to 30 ◆ Transport – M25 cuts across the top and M11 to the east ◆ Building pressure, especially in the southern sections ◆ Escapes from gardens allow new species to enter the ecosystem ◆ Fly tipping at night

Case study: Epping Forest *(continued)*

Official responses	◆ Corporation of London bought a further 720 ha to act as a buffer, especially in the north ◆ Byelaws passed to prohibit cycling and horse riding in sensitive areas ◆ Licensing – horse riders need a licence from the Woodland Authority ◆ Managed by conservators with a team of 80 staff to monitor the area ◆ SSSI status (1728 ha) and Special Area of Conservation (1605 ha) ◆ Epping Forest Management Plan (2004–10) set up entry gateways and visitor hubs to channel visitors and restored pollarding (75 ha), extended wood grazing (800 ha), cut grass to remove invaders, restored 8 ponds and created 3 new ones

Remember

This section looks at the direct and indirect threats posed to ecosystems but if asked to evaluate the relative level of threat you can also look at the positive impacts of human activities.

Exam tips

Epping Forest is a good example of a case study where a number of pressures (e.g. wildlife and recreation) compete for the woodland. This can lead to conflicts.

2.4 Why does the impact of human activity on the physical environment vary over time and location?

Student book page 67–72

How does the impact of human activity vary over time and location?

With time

- Level of development
- Short term versus long term
- Changes in technology
- Advancement of knowledge and understanding
- Changes in number and type of population
- Changes in cultural attitudes and priorities.

With location approach 1

- Development differs
- Technology differs
- Population differs (number and type)
- Incomes and wealth differ
- Cultures differ
- Political attitudes differ.

With location approach 2

- Highland versus lowland
- Coastal versus inland
- Core versus periphery.

Exam tips

The impact of human activity can be viewed as negative, neutral (conservation) or positive. It can also be seen as quantitative and qualitative (as a whole) or according to its impact on individual components of the environment.

Why does the impact vary with the type of development?

Economic development (Rostow model)

Environments are exploited initially, often as sources of raw materials to aid economic take-off. Industrialisation then damages the environment through pollution, sprawl, etc, but by the time there is high mass consumption there are also 'spare' funds to care for the environment.

Social development (Demographic transition)

Initially, population growth is rapid as death rates fall, so there is great pressure on the environment. But as birth rates then start to fall, growth slows so environments are put under less pressure. This reflects the public's attitude to the environment and its protection.

Technological development

Initially, technology is in balance with the environment but as technology develops there is greater capacity to exploit the environment on a larger scale (e.g. the mechanisation of farming), so harming it. With time, technology is developed to help protect and restore the environment. You could use this to illustrate the Boserup argument – that population pressure triggers technological change and innovation.

Key words

Take-off

Demographic transition

High mass consumption

Technology

Case studies: forests in Indonesia and the UK

	Indonesia **Student book pages 70–72**	UK **Student book pages 67–68**
	LEDC – preconditions to take-off	MEDC – high mass consumption
Type	Tropical rainforest	Temperate deciduous woodland
Impact of human activities	Increasing since 1960s (rainforest decreased by 40% between 1960s and 2005)	Decreasing since 1960s (forest area increased 3%; now 12%)
Reasons	◆ Increased foreign earnings from felling (timber is a major export) ◆ Oil palm plantations increased eight-fold in 20 years, at expense of forest ◆ Increased clearance for mining and oil exploitation ◆ Pollution from mining (e.g. mercury and cyanide used in gold mining) ◆ Growing population (1% each year) needing more land for food crops ◆ Settlement sprawl and road building ◆ Use of wood as fuel (no cheap alternatives) ◆ Illegal or unplanned burning of forest to clear area ◆ Country was slow to see conservation as important	◆ Seen as a way of developing rural areas (EU grants) ◆ Trees planted to regenerate waste areas (e.g. old coal tips) ◆ Increased use for recreation (e.g. new National Forest in Midlands, 520 km^2) ◆ Conservation – SSSIs and nature reserves (e.g. New Forest) ◆ Forests seen as source of renewable energy and a carbon trap ◆ Desire to landscape and add beauty (e.g. National Urban Forestry Unit created) ◆ Community Forest Scheme 1990 – planted 10 000 ha and upgraded 27 000 ha (12 community areas aiming to increase forests to 30% of each area) ◆ Tax incentives and grants for private planting (50% still private) ◆ Forestry Commission set up to manage forests and create jobs

Quick check questions

1 Currently a known species becomes extinct every how many minutes?

2 What prevented Epping Forest being built on as London expanded?

3 What is pollarding?

4 Why is conservation often regarded as a 'luxury'?

5 Why do tropical rainforests have such a great diversity of species?

6 Why are MEDCs increasingly seeing woodlands as sources of fuel?

7 Why is Indonesia clearing so much of its rainforest?

2.5 How can physical environments be managed to ensure sustainability?

Student book pages 72–78

Sustainability means ensuring that something exists in the future in a viable form.

Exam tips

Is it actually possible to create a sustainable environment? This is an issue that you should always debate in your answer. In most cases attempts are made to make an environment **more sustainable** rather than **absolutely sustainable**.

There is always some tension between economic and environmental sustainability. The compromise may be sustainable development (e.g. ecotourism, sustainable forestry, etc).

Strategies used include:

- Restricting access – rationing entry, closing off certain areas, etc
- Zoning of an area for particular activities – honeypot sites to attract the bulk of the pressure (e.g. from tourism)
- Visitor codes of conduct – education, regulations, laws, etc
- Planning controls to minimise environmental impacts
- Conservation – tight protection of an area
- Sustainable development.

UK sustainability initiatives

Individual initiatives

- Landowners – e.g. the Rothiemurchus Estate in the Cairngorms, Scotland, one of the largest remaining areas of Ancient Caledonian Forest, has a wide range of habitats with many rare species including capercaillie, pine marten, over 125 bird species, 27 animal species and 400 species of flowers.
- Farmers – e.g. the Farmers Conservation Group is a national network of farmers' groups that promotes the integration of efficient farming with game and wildlife conservation. The national membership now stands at almost 3000 farmers, all of whom have an interest in managing farmland habitats for wildlife.

Key words

Honeypots
Zoning
Carrying capacity

Corporate initiatives

- Private – large companies may re-instate or protect ecosystems on their sites e.g. Fawley Oil Refinery has a salt-marsh SSSI.
- Military – often manage the environment while managing military ranges (e.g. Bovingdon Heath in Dorset).
- Charities – the National Trust manages e.g. Wicken Fen in Cambridgeshire (has open fen habitats such as sedge beds, reed communities and fen meadows) and the RSPB manages e.g. Minsmere, Suffolk (with nature trails through woodland, heathland and reedbeds, as well as hides and access to the beach).

Local authority initiatives

- Planning controls (e.g. green belts, parks, tree preservation orders)
- Can designate local nature reserves.

National government initiatives

- National Park policies
- Over 4000 SSSIs in England, covering 8 per cent of area (e.g. limestone pavements)
- Over 220 NNRs in England, covering nearly 1 per cent of the area (e.g. Dawlish Warren, Devon, contains the full range of coastal habitats, from mudflats to sand dunes and up to 8000 wading birds)
- There are 36 AONBs in England, covering 15 per cent of the area (e.g. the Cotswolds, which is the largest, totalling 2038 sq km^2).

EU initiatives

- The Habitats Directive May 1992 requires EU member states to create a network of protected wildlife areas, known as Natura 2000, across the European Union. This network consists of Special Areas of Conservation (SACs) and Special Protection Areas (SPAs), established to protect wild birds. Currently covers 850 000 km^2, representing more than 20 per cent of total EU territory, e.g. Breckland on Norfolk/Suffolk border is an SAC of 7548 ha.

Key words

AONBs (Areas of Outstanding Natural Beauty)
NNRs (National Nature Reserves)

Exam tips

It is worth considering why ecosystems are increasingly managed at an international as well as local level. The size needed to make an environment self-sustaining often means that it crosses national borders.

International sustainability initiatives

International Convention on Biological Diversity 1992

This convention recognises that biodiversity is a 'common concern of humankind'. It covers all ecosystems, species and genetic resources and has three aims:

1. conservation of biodiversity
2. sustainable use of the global ecosystem's components
3. fair and equitable sharing of genetic resources.

Currently 188 countries and the EU have signed up to this convention. It focuses on ecosystems and looks at how to protect areas and ensure biosafety, reduce invasive alien species, and gain access to and share the benefits of genetic resources. The convention is administered from Montreal, Canada.

Global Strategy for Plant Conservation 2002

This global strategy aims to slow the global rate of plant extinctions by 2010. So far, 191 countries have ratified the convention, which has six aims including:

- establish conservation status of all known plants
- protect 50 per cent of the most important species
- conserve 60 per cent of threatened plant species
- ensure that at least 30 per cent of plant-based products come from sustainable sources.

Exam tips

Think about the penalties of not following strategies, having agreed to them. Are there any realistic incentives to encourage countries to follow international conventions?

Case study: The Great Barrier Reef

Student book pages 73–74

Background

- Largest coral reef in the world – 2300 km long
- Supports over 1500 types of fish and 4000 types of mollusc
- Attracts 2 million visitors and 5 million recreational users a year
- Tourism generates AU$5 billion a year and 54 000 jobs.

Threats

- Agricultural runoff – eutrophication
- Removal of mangrove swamps along coast has led to silting
- Sewage effluent from coastal settlement
- Collecting of coral by divers and increasing tourism
- Erosion from boats and visitors
- Crown of thorns starfish is eating the reef
- Global warming (coral bleaching).

Management

- Great Barrier Reef Marine Park Authority (GBRMPA) created in 1975
- Charges visitors to the reef a AU$4.50 daily fee
- Acts as planning authority for whole of reef
- Reef divided into seven zones for different activities (e.g. conservation zone)
- Fines for breaching regulations and patrols by wardens
- Honeypot sites created but boat size limited and visitor numbers controlled
- Education and various signs to warn of sensitive areas
- Fishing is controlled and limited
- Many species are legally protected
- Made a World Heritage Site in 1981.

Disadvantages and benefits of conservation

Disadvantages of conservation

- Conservation can displace or disrupt local communities (e.g. the Masai in the Masai Mara Reserve in Kenya).
- It may over-protect (by removing natural threats), e.g. prevention of fires led to loss of species needing fire to regenerate in the Smoky Mountains National Park, USA.
- It prevents natural change. Ecosystems and environments are dynamic systems so they should be allowed to change.

Benefits of protecting biodiversity

- Many species contain as yet unrecognised resources for humans e.g. medicines (currently 50 per cent come from natural sources and 80 per cent of these from tropical rainforests).
- Vegetation helps maintain air quality and supplies oxygen.
- Plants absorb atmospheric carbon so reducing the greenhouse effect.
- Many natural environments provide native populations with resources, e.g. fuel and food.
- Water purification. Vegetation filters water and wetlands are particularly effective at removing nitrogen from water.
- Vegetation reduces flooding by absorbing runoff or slowing it. In coastal areas it can absorb wave energy and reduce coastal flooding.
- Reducing the genetic pool leads to an increase in disease. It is difficult to find species that are disease-resistant, e.g. two species of wheat account for 99 per cent of our wheat crops.
- Maintains the health of the soil by having the right balance of nutrients and soil flora and fauna. This reduces soil erosion.
- Aesthetics: the natural world is often beautiful and helps reduce people's stress levels.

Quick check questions

1. Why would farmers want to protect wildlife on their farms?
2. What is the main advantage for coastal refineries of protecting the salt-marshes?
3. Why do military ranges make such good refuges for wildlife?
4. Why is direct ownership (e.g. by The National Trust) the most effective way of protecting an area?
5. What do you think is the most important argument for maintaining biodiversity?
6. Besides coral bleaching, why is global warming a major threat to the Great Barrier Reef?

ExamCafé

Student book pages 46–83

Section A

Sample question

Study resource 2, a diagram of the trophic levels in an ecosystem where there appear to be excessive carnivores.

Identify **one** *issue and suggest appropriate management strategies to deal with it. [10 marks]*

This is one of those examples where there is only one obvious issue and it is clear to see. More stretching is the requirement to suggest appropriate management strategies, as this will really show whether you have understood the concept of the trophic levels and the relationship between them.

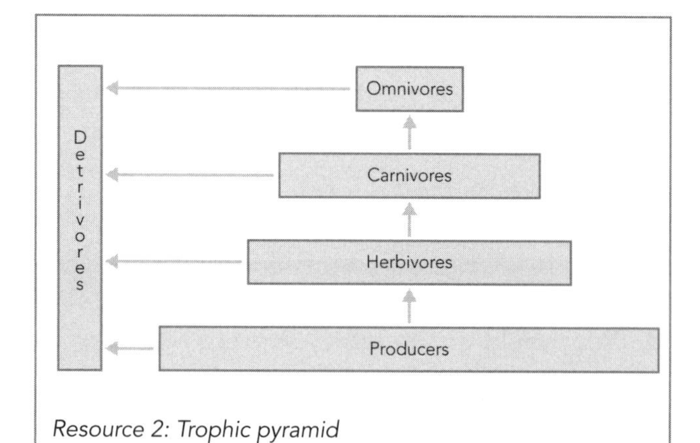

Resource 2: Trophic pyramid

Student Answer

This can be easily adjusted by hunting the excessive carnivores and so bringing the system back into balance.

Clearly this is an inadequate answer for the following reasons:

- How can 'excessive' carnivores be identified?
- Hunting is often selective, removing the older or weaker carnivores.
- Would birth control be a fairer method, or increasing herbivore numbers?
- This management strategy ignores the knock-on effects on the levels above and below. Excessive hunting could lead to an explosion of herbivores and their subsequent starvation when they overgraze the vegetation.

Would an alternative answer be to do nothing as the system would eventually adjust itself, with the 'excessive' number of predators dying off because they have little to eat? Doing nothing is indeed a strategy but this answer would need to be supported with some details of careful research and planning to avoid excessive suffering. Clearly it would be crucial to know the size and nature of the area involved. Although you would not be expected to know such details from a simple resource, you could be expected to appreciate their significance in deciding on an appropriate strategy.

Section B

Sample question

For one named local ecosystem or environment assess the relative threats to it from human activity. [30 marks]

The focus is on section 3 of the specification and there is a clear requirement to study one local ecosystem or environment. 'Local' is a scale as much as a locational directive so it does not have to be local to you and may have been an example studied during fieldwork. The specification quotes some examples of local ecosystems or environments: woodland, dunes or a marsh (salt or freshwater).

The first step is therefore to locate your local ecosystem and show that it is a real place. The best solution is to draw a sketch map to locate it and show some of its major or significant characteristics. Always give some indication of scale and north but there is no need to use colours as these will not show up on the scanner because your paper will be marked online.

Then you need to think about how to structure your response. You will need to make some reference to the scale, timing and nature of the threats.

Student Answer

The chief threat to Belfairs Wood probably involves the biomass store as coppicing removes some of the store, and the trampling of the ground by numerous visitors impedes the take-up of nutrients and water by the trees.

This answer is not entirely correct but the candidate did try to identify the parts of the system under threat and so demonstrate an understanding of nutrient stores and flows. The stress is on evaluating the relative threat of human activities such as recreation, farming, industry, settlement, transport, etc. Even conservation or biodiversity protection can be a threat to the free functioning of the ecosystem!

Another answer would be:

Student Answer

The chief threat to Belfairs Woods comes from recreation. Other threats are reduced as the area is a nature reserve, which restricts its development and prevents the encroachment of building or road development.

There is nothing wrong in explaining why some activities pose little or no threat (that is, evaluating) but it is important not to ignore larger threats, such as global warming, pollution, etc, which pose problems for all environments. Quality answers will consider why the relative importance of the threats from a range of activities may vary with time (over a year or longer periods) or according to which aspects of the ecosystem they affect. It is important not to get sidetracked into a discussion that is not strictly relevant to the question such as ways of reducing or modifying these threats.

Remember: your conclusion should draw these threads together and give a clear assessment of the major threats to that particular ecosystem and why you think they are the important ones.

Chapter 3 Climatic hazards

Student book pages 84–123

3.1 What conditions lead to tropical storms and tornadoes and in what ways do they represent a hazard to people?

Student book pages 86–95

Causes and effects of tropical storms and hurricanes

Remember

There is a difference between tropical storms and hurricanes. The latter is a regional name. From an examination point of view, you can treat them as the same.

Definitions

- Tropical storms: winds of 64–118 km/h
- Hurricanes: winds of over 118 km/h

Conditions for

- Intense tropical depressions
- Seasonal: June–November in northern hemisphere.

Formation

- High humidity releases latent heat
- Sea temperature of more than 26–27°C for at least 60 m depth
- More than 5° north or south of Equator for Coriolis effect to impart spin
- Almost constant vertical conditions
- Divergent airflow with height to draw air upwards
- Unstable air – surface winds converge
- Two moist tropical airstreams meet and denser one undercuts the other.

Development

- Move westward due to Earth's rotation – 15–30 km/h
- Continue to grow as condensation releases heat energy and picks up warmth (energy) from the sea
- Break up over land (friction and little moisture) or as they move out of tropics 35°N/S as sea gets cooler.

Characteristics

- Central clear 'eye' of descending air
- Around eye, massive cloud walls with rapid uplift
- 12 km high, 200–500 km in diameter
- High volume rainfall – up to 500 mm in 24 hrs
- Very low air pressure – 900 mb
- Eye wall dominated by intense thunderstorms
- Movement difficult to predict.

Impacts from

- Wind, storm surge (low pressure), intense high rainfall, lightning.

Types of impacts

- Primary – homes destroyed, deaths and injuries from flying debris
- Secondary – flooding, pollution (sewage/drains, etc), disease, hunger, fires (power lines down), transport disrupted
- Tertiary – long-term economic impacts (e.g. cost, destruction of infrastructure, loss of jobs, etc).

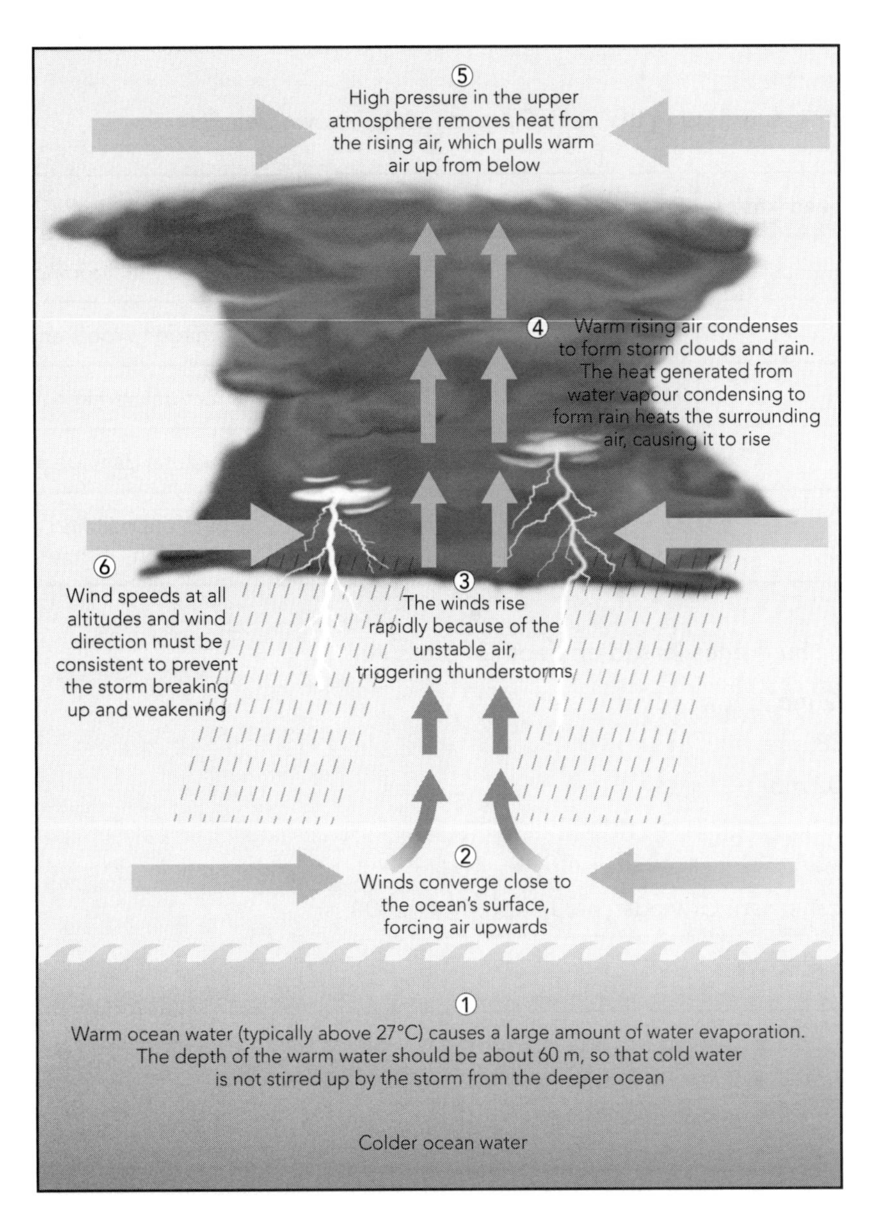

Formation of a hurricane

Environmental impacts

- Relief – landslides, mudflows
- Drainage – floods, waterlogging
- Vegetation – trees destroyed/uprooted, habitats destroyed
- Pollution of water supplies – disease.

Social impacts

- Health – injuries and deaths, disease, depression
- Housing – destroyed, temporary shelter, forced to migrate
- Social unrest – looting, family break-up, tension (e.g. Hurricane Katrina).

Economic impacts

- Infrastructure – destroyed (e.g. roads, power, schools)
- Agriculture – cash and food crops lost, pollution, tree crops hard hit
- Transport – bridges destroyed, road and rail damage, loss of aeroplanes
- Trade – loss of exports, need to import, cost of aid.

Exam tips

Be careful, as it is easy to confuse primary/secondary hazards with primary/secondary impacts. Read the question carefully to check which is needed. Hazards are the potential dangers from an event; impacts are the specific effects of a particular event. Primary hazards are therefore wind, rain and high tides, while secondary hazards could include flooding, landslides and disease.

Key words

Eye	Eye wall
Coriolis effect	Storm surge

Hurricanes

Hurricane strength is measured using the Saffir-Simpson scale, shown below:

Category	Windspeed (km/h)	Pressure (mb)	Storm surge (m)	Damage
1	120–53	980+	1.5	Limited but flooding
2	154–77	965–79	2.0	Damage to roofs and vegetation
3	178–209	945–64	3.0	Structural damage and storm surge
4	210–49	920–44	4.5	Structural damage and mass evacuation
5	250+	<920	>5.5	Complete building failure and mass evacuation

Some extreme Atlantic hurricanes (based on depth of pressure):

- Wilma, 2005 – 882 mb
- Gilbert, 1988 – 888 mb
- Katrina, 2005 – 902 mb

Case study: Hurricane Ivan, Grenada, 5–20 September 2004

Student book pages 90–91

Characteristics	◆ Winds reached 270 km/h ◆ Pressure 902 mb ◆ Relatively little rain ◆ Generated 117 tornadoes ◆ Massive waves reaching 40 m but little storm surge ◆ Led to the deaths of 121 people and damage of US$17 billion ($13 billion in the USA)
Impacts in Grenada	
Primary	◆ 39 killed ◆ 85% of island devastated ◆ 90% of homes damaged or destroyed ◆ 50% of population made homeless
Secondary	◆ Looting cost US$800 million ◆ Roads blocked by debris ◆ Local flooding due to blocked drains ◆ Loss of phone lines ◆ Loss of power as lines down
Tertiary	◆ Damage cost US$1.1 billion ◆ Loss of export cash tree crops, e.g. nutmeg ◆ GDP growth was projected at 4.7% for 2004 but it actually fell by 3% ◆ Massive increase in Grenada's national debt (may have to default on its loans) ◆ US$150 million in aid sent to the country ◆ Loss of growing tourism industry (60% of jobs in tourism lost)

But Hurricane Ivan could have been worse. As there was little rain, there were no mudslides, the storm surge was minor and it occurred during the day.

Exam tips

Hurricane Ivan is a good case study for a hurricane in a LEDC and could well be compared with Hurricane Katrina in a MEDC (see pages 39–41).

Tornadoes

Definition

- Violent rotating small-scale wind storms – can reach 500 km/h

Conditions for formation

- Seasonal: June–November in northern hemisphere
- Often a remnant of a tropical storm.

Formation

- Air masses of differing temperature and humidity meet
- Anvil thunderclouds, known as supercells, exist
- Downward rapid current of cool air
- Almost constant vertical conditions
- Divergent airflow with height to draw air upwards
- Very unstable air – surface winds converge or very hot surfaces
- Flat land, as easily disrupted by relief.

Development

- Continue to grow as condensation releases latent heat energy
- Break up over land (due to friction and little moisture) or where land is cooler.

Characteristics

- Small and short lived (rarely over an hour) but highly destructive
- Elongated funnel of cloud (vortex) – in contact with cloud and ground
- 200 m in diameter
- High volume rainfall
- Twisting wind rotation so lifts objects
- Very steep pressure gradient (25 mb per 100 m) – centre is at very low pressure. Pressure difference so great that buildings 'explode'
- Movement difficult to predict but tend to follow certain routes (e.g. tornado alley in Kansas).

Tornado strength is measured using the Fujita scale, shown below:

Category	Wind speed km/h	Damage
0	<117	Light (e.g. trees blown over)
1	117–180	Moderate (e.g. moving cars blown over)
2	181–253	Considerable (e.g. roofs torn off)
3	254–332	Severe (e.g. parked cars lifted)
4	333–418	Devastating (e.g. houses levelled)
5	419–512	Incredible (e.g. houses blown away)

Tornadoes have serious environmental, social and economic impacts.

Types of impacts

- Primary – destruction of buildings, crops, deaths
- Secondary – cost of rescue, loss of power, depression
- Tertiary – long-term economic impacts, e.g. cost, loss of jobs, etc.

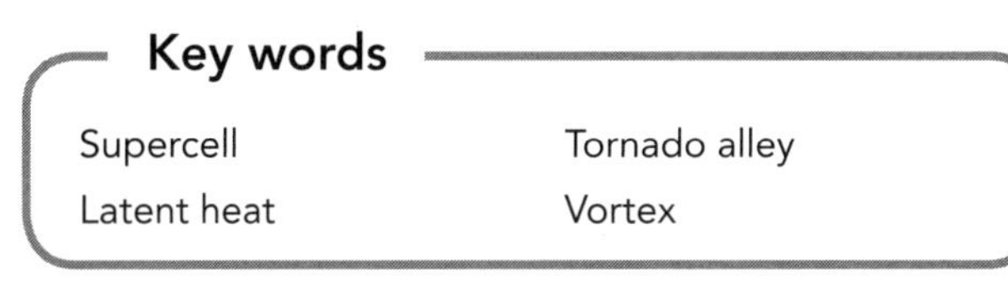
Key words

Supercell
Tornado alley
Latent heat
Vortex

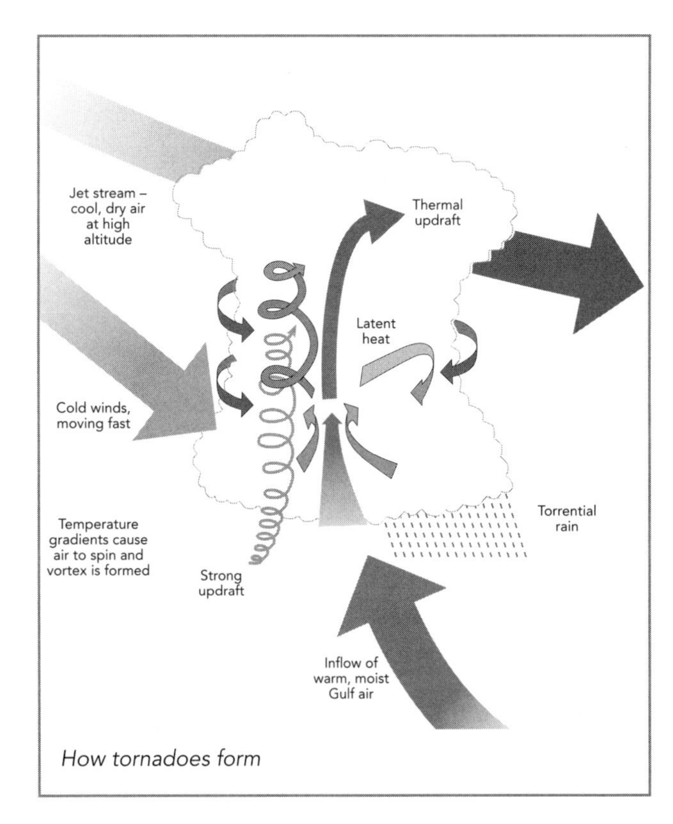

How tornadoes form

Environmental impacts

- Drainage – choked with wind-blown debris
- Vegetation – trees destroyed/uprooted, habitats destroyed
- Pollution of water supplies – disease.

Social impacts

- Health – injuries and deaths, disease, depression
- Housing – destroyed, temporary shelter
- Social unrest, looting.

Economic impacts

- Infrastructure – destroyed, e.g. roads, power, schools
- Agriculture – cash and food crops lost, pollution, tree crops hard hit
- Transport – bridges destroyed, road and rail damage, loss of aeroplanes
- Cost of aid
- Industrial capacity reduced – damaged buildings, loss of power, etc.

Remember

The country with the largest number of tornadoes per unit area is the UK, yet few do much damage.

Case study: Indiana tornado, November 2005

Characteristics	◆ Developed along a squall line along a cold front ◆ 4 tornadoes formed from 2 supercells ◆ Cut a swathe of damage 400 m wide and 66 km long ◆ Wind speeds reached 320 km/h ◆ It lasted 10 hours ◆ Unusual, as it occurred in November (most are March to June) and at 1.50 am (most occur during the day).
Impacts	◆ 25 killed and 230 injured ◆ US$92 million of damage ◆ Gas leaks ◆ 25 000 left without power ◆ 225 mobile homes destroyed
Responses	◆ Rescue services were on site very quickly, as they were prepared ◆ US$2.4 million of state aid made quickly available for housing, etc ◆ Long-term grants for rebuilding

Quick check questions

1 What latitudes do hurricanes a) start in b) die out above?
2 Why do hurricanes increase in velocity over the sea?
3 What is the chief secondary impact of a hurricane?
4 What scale is used for hurricane strength?
5 Why do most deaths occur after the eye has passed overhead?
6 Why do buildings 'explode' in a tornado?
7 What was unusual about the 2005 Indiana tornado?
8 Why do tornadoes cause less damage than hurricanes?

3.2 How do atmospheric systems cause heavy snowfall, intense cold spells, heatwaves and droughts, and in what ways do they represent a hazard to people?

Student book pages 96–105

A) High pressure systems: anticyclones

Characteristics

- Large area of slow-moving air
- High pressure, as air is sinking
- As air sinks it warms, causing temperature inversion
- As air is sinking it can't rise to form clouds so there are clear skies
- Winds are light, as gentle pressure gradient
- Winds blow outwards, usually clockwise, in the northern hemisphere
- Usually ends when uplift overcomes temperature inversion, leading to thunderstorms.

Weather associated with high pressure		
	Summer	Winter
Temperature	◆ Hot in the day ◆ Chilly at night, as no cloud to trap heat	◆ Cold in day ◆ Freezing at night = penetrating frost ◆ Cold spells
Cloud	Little but cumulus may form during the day	Little
Precipitation	Low – drought but may end with a heavy thunderstorm	Very low, as so cold, but if it warms snow can occur
Wind	Little – mostly calm	Little – calm but if it blows then intense wind chill
Sunshine	Long hours of sunshine = heat stroke	Long hours of sunshine but if fog gets trapped under inversion layer it leads to anticyclonic gloom
Humidity	Varies with source of air, e.g. if tropical then very humid	Low = very dry conditions
Visibility	Mist in the early morning or over the colder sea, combined with heat haze, reduces visibility	Fog common, especially along coasts near warmer sea or collecting in hollows

B) Low pressure systems: depressions

These develop along the polar front where cold northerly air meets/undercuts warm tropical air moving north. Depressions form as waves along this front. (The polar front occurs in both northern and southern hemispheres and moves with the seasons.)

Characteristics

- Relatively small area of fast-moving air
- Low pressure, as air is rising
- Tend to move west to east across the UK
- As air is rising it forms clouds, so bringing rain
- Winds are strong, as steep pressure gradient
- Winds blow inwards, usually anticlockwise, in the northern hemisphere
- Rarely last more than a day but frequently one in a stream of depressions
- Depressions have cold and warm fronts, as the warm air is undercut by the cold.

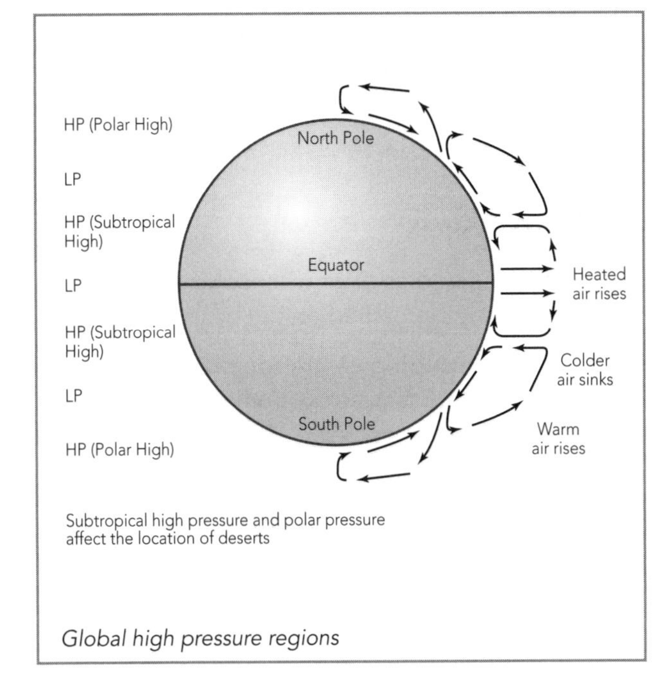

Global high pressure regions

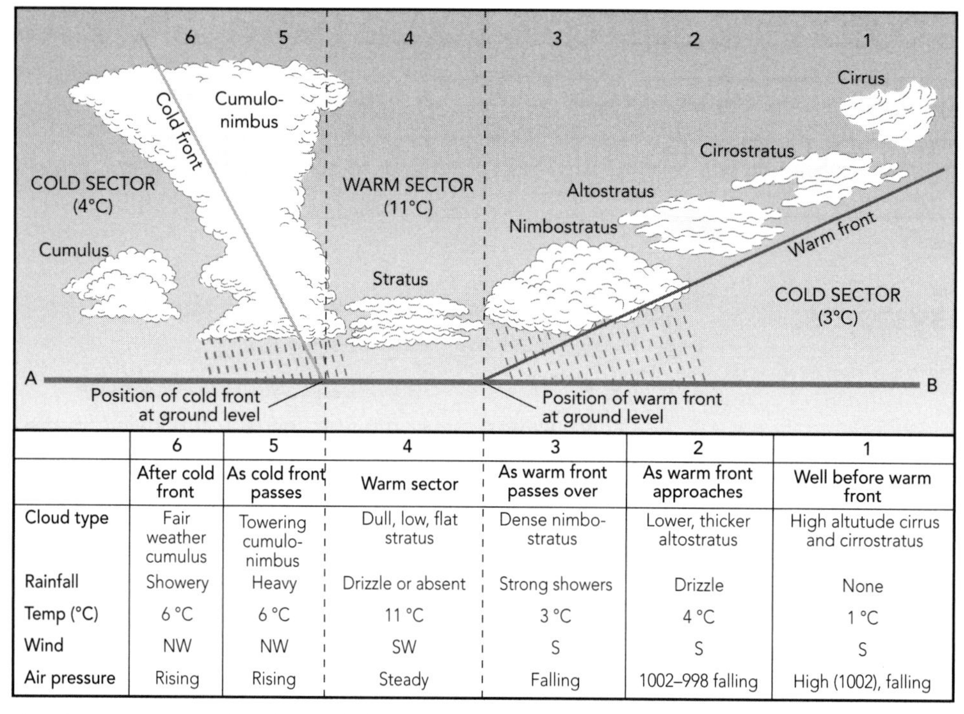

	6	5	4	3	2	1
	After cold front	As cold front passes	Warm sector	As warm front passes over	As warm front approaches	Well before warm front
Cloud type	Fair weather cumulus	Towering cumulo-nimbus	Dull, low, flat stratus	Dense nimbo-stratus	Lower, thicker altostratus	High altutude cirrus and cirrostratus
Rainfall	Showery	Heavy	Drizzle or absent	Strong showers	Drizzle	None
Temp (°C)	6 °C	6 °C	11 °C	3 °C	4 °C	1 °C
Wind	NW	NW	SW	S	S	S
Air pressure	Rising	Rising	Steady	Falling	1002–998 falling	High (1002), falling

Weather associated with a 'typical' depression

Source: Guinness, P. and Nagle, G. AS Geography: Concepts and Cases (2000), Hodder Murray, reprinted by permission of John Murray (Publishers) Ltd.

Weather associated with low pressure (very similar, winter or summer)		
	Summer	Winter
Temperature	Mild in the day Warm at night as cloud traps in heat	Mild in the day Warm at night as cloud traps in heat. No frost
Cloud	Heavy especially at the fronts	Heavy especially at the fronts
Precipitation	High – often heavy thunderstorm producing intense downpours	High – usually rain but can be snow over hills. With strong winds this causes blizzards
Wind	Strong and gusty	Strong and gusty so can cause wind chill
Sunshine	Relatively little sunshine	Little sunshine – very dull and gloomy
Humidity	High as warm air holds a lot of moisture	High as warm air holds a lot of moisture
Visibility	Good but poor in rain and can cause hill fog as clouds are level with summits. Also frontal fog where air masses meet	Good but hill fog common

C) Air masses

These are large bodies of air with fairly uniform conditions. They are named on the basis of:

- their origin in high-pressure source areas: polar (from high latitudes), arctic (from the Arctic), tropical (from the tropics)
- what they have travelled over: maritime (ocean), continental (land).

Key words

Temperature inversion
Precipitation
Radiation fog
Anticyclonic gloom

Exam tips

The most effective way to show the weather in a depression is to use a cross-section diagram.

Air masses influencing the climate of the British Isles		
Air mass	**At origin and modification**	**Weather for the British Isles**
Polar maritime (from north-west and west)	Cool, dry and stable but picks up moisture over warm sea and becomes more unstable	◆ The dominant air mass – 80% ◆ Cloudy, cool, wet weather
Tropical maritime (from south-west)	Warm, moist and unstable but becomes more stable as it moves north over cooler sea	Tends to occur in summer, very humid and if forced to rise brings thunderstorms
Tropical continental (from south)	Warm, dry and unstable and rarely changes as it moves north in summer over warm dry land	◆ Brings poor visibility (Saharan dust) and fog over sea but little rain ◆ Very hot weather
Polar continental (from east)	◆ Cold, dry and stable and this increases as it crosses cold land in winter ◆ If it comes from north-east it crosses the North Sea, which warms it and adds moisture	◆ Brings bitter cold winds, causing very low temperatures ◆ If it crosses the North Sea it can bring heavy snow to east coast and fog over sea
Arctic maritime (from north)	Very cold, dry and stable but becomes wetter and less stable as it moves south over sea	Brings heavy snowfalls and bitter wind chill

Key words

Stable = where air, once forced up, tends to sink down again

Unstable = where air, once forced up, tends to keep on rising

Remember

The UK's weather is dominated by the interaction of anticyclones and depressions, which pull in different air masses.

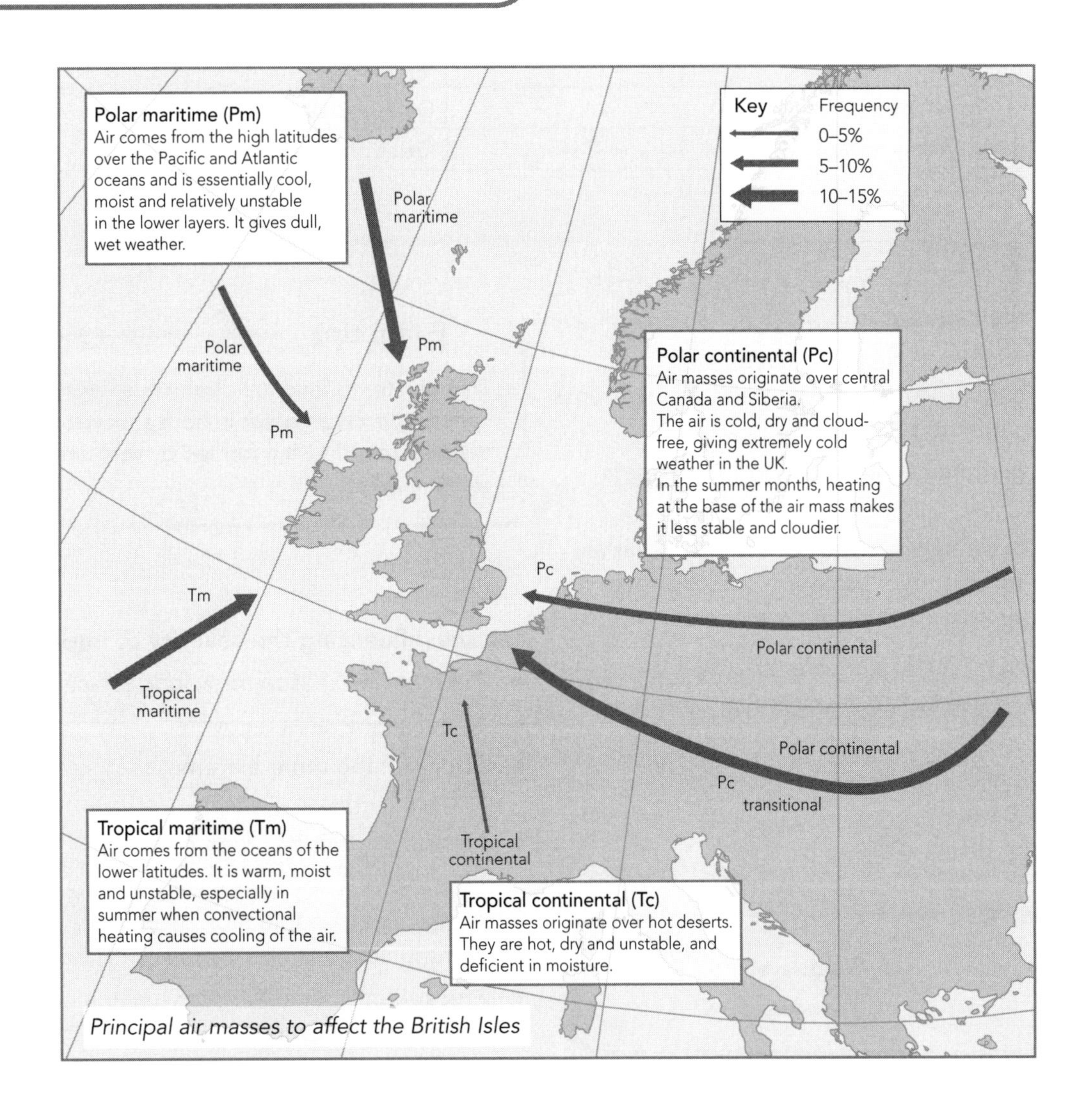

Principal air masses to affect the British Isles

What hazard impacts do high and low pressure systems bring?

	Anticyclones	Depressions
Environmental		
Ecosystems	Drought, fires, heat, frost	Flooding, gales and storms
Vegetation	◆ Trees susceptible to drought (lose their leaves) ◆ Frost kills seedlings and buds (can't reproduce) ◆ Trees wilt in heat, as increased transpiration means they need more water	◆ Lack of sun ◆ Trees blown down
Drainage	◆ Bakes ground in drought so flash floods when it rains ◆ Rivers and lakes dry up	Flooding, waterlogging
Soils	Dry and shrink = subsidence	Swell = landslides, creep, soil erosion
Social		
Health	◆ Poor due to heat/cold/fog and high pollen count ◆ Heat stroke	Dampness = bronchitis and depression
Housing	Heat and cold damage the fabric	Wind can damage housing (e.g. roofs)
Accidents	Frost and fog cause poor driving conditions	◆ Wet roads and flooding ◆ Gales
Economic		
Agriculture	Need to irrigate, frost kills crops	Lack of sun, waterlogging of crops, wind damage
Forestry	Fires, droughts	Waterlogging, gales
Transport	Accidents in fog and on frost, heat buckles rails and melts roads	◆ Wet surfaces – aquaplaning ◆ Strong winds halt flights and shut bridges
Industry	Water shortage	Gales and flooding
Power	Water shortage	Gales bring down power lines

Hazards and their causes

- Heavy snowfall and blizzards = arctic maritime and polar continental North Sea track, depression following prolonged anticyclone in winter
- Frost = polar continental, anticyclone in winter, cold spells
- Fog = anticyclones especially in autumn, polar air masses in summer over the warm land and over sea in winter, tropical over sea in summer
- Drought = tropical continental, anticyclone
- Heatwaves = tropical air mass, anticyclone in summer
- Thunderstorms and heavy rain = tropical maritime, depression
- Gales = depression

Key words

Drought
Heatwave
Blizzard
Thunderstorms

Exam tips

Thunderstorms (and their hazards of high rainfall, lightning, etc) can occur in both anticyclones and depressions. They are formed by very unstable air rising rapidly.

Factors influencing the severity of impacts

- Time of day/season, how long it lasts
- Scale: size of area it covers
- Strength: the more extreme, the greater the impact
- Frequency: is it common or rare and thus unexpected?
- Awareness of population: warnings, communications, are people prepared?
- Precautions taken: long-term planning and short-term measures.

Case studies: North American blizzard and European drought

	North American blizzard, February 2003 **Student book page 101**	European drought, 2003 **Student book pages 104–05**
	◆ Hit the east coast, especially Boston to Washington ◆ Lasted 5 days	◆ A drought is 15 consecutive days without rain ◆ The 2003 drought lasted from February to October
Cause	A blocking high pressure system over Canada forced air from Rockies (PC air mass) to cross the cold interior but then it met the moist warmer air of the coast	A series of intense anticyclones (blocking highs) that forced depressions north or south of the UK
Nature of the hazards	◆ Snow 40–80 cm deep ◆ Very low temperature and wind chill ◆ 75 km/h winds ◆ Blizzards (in Boston 70 cm snow fell but massive drifts)	◆ Lowest rainfall since 1921 and in August–October 77% below average ◆ Evaporation 15% above normal ◆ High temperatures (37°C recorded) ◆ Thunderstorms and electrical storms, as the air was so dry
Impacts	◆ 27 deaths (car and other accidents and elderly deaths) ◆ Transport paralysed – airports, rail and road links shut ◆ Major power cuts as lines brought down – 95 000 homes blacked out in West Virginia ◆ Schools closed for a week ◆ Some damage to buildings from weight of snow – cost US$14 million ◆ High cost of clearing snow (e.g. US$20 million in New York)	◆ High levels of pollution, high pollen count (hayfever) and danger of sunstroke and skin cancer ◆ Expansion in tourism (4%+), mainly to the cooler coast ◆ Retailers expanded sales of beer, ice cream, etc ◆ 900 deaths from poor air quality (high levels of ozone) ◆ Loss of work days – people took time off (estimated cost of £10 million per day) ◆ Harvest yields fell 20%, and milk yield fell 15% ◆ Water shortage so hosepipe bans ◆ Roads melted in Essex, rails buckled ◆ Subsidence in buildings as ground dried up and shrank ◆ Trees hard hit – wilted and died ◆ Reservoirs dried up (50% below capacity) – standpipes needed in driest areas ◆ Fires, as the land was so dry (e.g. rare birds wiped out on Dorset heaths) ◆ Water-using industries were hard hit, e.g. swimming pools, golf ◆ Increased eutrophication (and fish deaths) in East Anglia

Remember

Whilst there is no requirement to know how these hazards are managed in this section, management of hazards could be referred to in section 3.4.

Quick check questions

1 Why do anticyclones contain temperature inversions?
2 Why are nights so cold in anticyclones?
3 Why is fog common in anticyclones?
4 In a depression, which front brings the heaviest rainfall?
5 What is the term used for the front formed when a cold front has caught up and merged with a warm front?
6 Why is the width of sea crossed by polar continental air masses such an important influence on the weather conditions in the UK?
7 Why does frost have such an impact on plants?
8 Why does the UK not prepare more thoroughly for heavy snowfalls?

3.3 Why do the impacts of climatic hazards vary over time and location?

Student book pages 105–08

Over time

- Level of development/technology
- Duration
- Recurrence interval and frequency of event
- Time of day/season/year
- Build-up of events – warning interval.

With location

- Coastal versus inland
- Ability to predict or forecast (level of technology)
- Population density, distribution, level of perception and education
- Level of communications – warning, mobility of population
- Highland versus lowland
- Urban versus rural
- Level of development – building type, ability to warn/evacuate
- Remoteness
- Type and size (power) of hazard or mix of hazards
- West versus east (depressions move west to east in the northern hemisphere) or north versus south (hurricanes die out as they move north and cool down).

Exam tips

Many questions focus on the contrast in the impact of weather systems on LEDCs compared to MEDCs or on contrasts in wealth or levels of development.

Case study A: LEDC, Smog in Indonesia 1997–98

Student book pages 105–07

Scale	◆ An area larger than western Europe including Malaysia, the Philippines and Thailand ◆ 70 million people affected
Causes	◆ Burning (often illegally) of forest and plantations to clear the land cheaply and easily, often organised by TNCs ◆ El Niño effect led to drought, which dried up the forests ◆ High pressure so inversion layer trapped smog ◆ Late arrival of monsoon so no rain to put out fires and clear the air ◆ Weak and corrupt local government made it difficult to enforce anti-burning laws ◆ Damage to rainforests made them easier to burn ◆ Lack of alternative fuels for the poor ◆ Inadequate fire-fighting resources
Impacts	
Short term	◆ Over 60 000 treated for smog-related illnesses and eye problems ◆ Airliner crashed, killing 234; flights cancelled or re-routed ◆ Ships re-routed to avoid Straits of Malacca ◆ Schools closed ◆ Loss of tourism ◆ Loss of expatriates
Long term	◆ Over 275 died from starvation ◆ Deaths from cholera due to lack of clean water ◆ Crop yields fell and many areas had to import food

Case study A: LEDC, Smog in Indonesia 1997–98 *(continued)*

	◆ Added to global warming and reduced carbon dioxide-absorbing vegetation ◆ Loss of biodiversity – species wiped out and food chain disrupted ◆ Peat may burn for 20 years so seeds can't germinate
Reactions	◆ Put out fires, e.g. using cloud seeding, water bombers (from USA) and fire fighters from Australia ◆ Stop fires being lit – companies found guilty of starting fires have their operating licences revoked ◆ Develop industries other than timber and tree crops (palm oil particularly) ◆ International cooperation – in 2000 ASEAN adopted a zero burning policy to solve the problem

Exam tips

Think about whether such reactions will effectively stop TNCs clearing the land, as they have so much economic and political influence in the area.

Key words

El Niño	TNCs (transnational corporations)
Smog	Biodiversity

Case study B: MEDC (urban area), Hurricane Katrina, August 2005, New Orleans

Scale	Hit an area the size of the UK
Causes	◆ Level 5 hurricane ◆ Storm surge of 5 m ◆ Winds of 195 km/h ◆ 49% of city below sea level – much of area subsiding ◆ Years of little maintenance work on protecting levees – complacency
Impacts	
Short term	◆ 1836 dead and 705 missing ◆ 50 breaches in levees flooded 80% of city, especially poor black areas ◆ 90% of population evacuated ◆ 3 million people lost power and phones ◆ Water pollution – sewage, chemical waste, etc ◆ Looting ◆ Loss of road and rail links ◆ High-rise buildings suffered wind damage
Long term	◆ Water- and mosquito-borne disease (e.g. West Nile Fever) ◆ Erosion of beaches and islands – 562 km^2 ◆ 20% of coastal marshes destroyed = loss of habitats ◆ 1 million people reluctant to return home – impact on evacuation areas ◆ 30 oil platforms and 9 refineries put out of production (20% of US capacity) = oil price rise ◆ 5300 km^2 of commercial forest lost ◆ Long-term damage to cotton and other crops, causing unemployment ◆ Higher insurance costs or people even unable to get insurance cover
Reaction	*See next page*

3.4 What can humans do to reduce the impact of climatic hazards?

Student book pages 108–12

Measures used to reduce impacts

1. **Prediction** – based on past events (hazard mapping), monitoring of pressure, satellites, weather balloons, etc
2. **Risk assessment** – calculate size and extent of risk, inform population, adapt building design and location of vital buildings/facilities (e.g. power stations)
3. **Prevention** – seed depressions/hurricanes, afforest slopes, raise and strengthen levees, build reservoirs, modify channels, move vulnerable population and livestock, establish exclusion zones, remove dangerous objects, e.g. signs
4. **Planning** – individual, e.g. store water, local authority (e.g. emergency centres or shelters), state or central (e.g. mobilisation of rescue services), building controls
5. **Preparation** – education (emergency drills), contingency plans (e.g. evacuation routes signposted, training of emergency staff, etc)
6. **Warnings** – use of media, communications, level of threat, planned evacuations
7. **Prevention** during event – dam or divert rivers, stabilise slopes, pass laws to force evacuation or water saving (e.g. hosepipe bans), rationing, etc
8. **Responses** – search and rescue, emergency aid, insurance, state or international aid for rescue and relocation
9. **Recovery** – clearance of debris, state aid for reconstruction, tax relief
10. **Redevelopment** – long-term plans.

Key words

Mitigation
Prediction
Risk assessment
Cost benefit

Exam tips

A very frequent question is one that asks you to compare the effectiveness of different types of strategies used to manage climatic hazards.

Case study: reducing the impact of Hurricane Katrina, 29 August 2005

Prediction	Proved very good but US National Oceanic and Atmospheric Administration (NOAA) is improving it by adding more weather buoys in the Gulf of Mexico
Risk assessment	Computer models gave a good assessment of areas at risk by 26 August
Prevention	Flood protection levees in place
Planning	◆ Local and state plans initiated by declaration of state of emergency on 27 August ◆ 57 emergency shelters opened and 31 extra shelters ◆ Plans followed several practice drills
Preparation	60 000 National Guard troops mobilised on 26 August, Coast Guard at the ready
Warnings	◆ Constant media warnings but mandatory evacuation order came only 19 hours before impact ◆ 80% evacuated but much of the transport infrastructure shut down
Responses	◆ Coast Guard rescued 35 000 people ◆ Federal Emergency Management Agency (FEMA) put in 28 search and rescue teams, supplied 85 million litres of water and 50 million meals, and rehoused 700 000 in trailers ◆ Red Cross supplied 68 million meals and shelter for 1.4 million families ◆ Salvation Army set up 178 canteen units and gave emotional and spiritual aid to 277 000 people
Recovery	◆ 100 m^3 of debris removed and 2414 km of channels cleared ◆ Over 1 million housed out of the area
Redevelopment	◆ US Army Corps has reinforced and raised 354 km of levees ◆ FEMA given US$5.5 billion to rebuild public infrastructure and government $17 billion for rebuilding

Despite all the impact-reducing measures, Hurricane Katrina ended up killing 1836 people and costing over US$150 billion.

Why did these measures fail?

- Lack of prevention strategies – levees poorly maintained
- Lack of coordinated planning between government agencies
- Delay in evacuation and reluctance of some to leave (often the poorest)
- Communications disrupted, especially loss of phones
- Sheer scale of the hurricane.

Exam tips

Hurricane Katrina is a good example of a MEDC not getting it right. Technology and wealth don't always buy reduced impacts.

Adjusting to hazards

Physical adjustments

- Altering characteristics of hazards (e.g. attempts have been made to seed clouds to reduce hurricane energy)
- Building to withstand hazards (e.g. hurricane shelters in Florida)
- Constructing diversions, barriers, etc (e.g. planting trees as windbreaks)
- Moving people to less vulnerable locations (e.g. away from low-lying areas of New Orleans).

Social adjustments

- Increasing public awareness via education, media, etc (e.g. tornado training for schoolchildren in the American Mid-West)
- Making local evacuation plans and preparations (e.g. use of radio and village communication systems in Bangladesh when a tropical storm approaches)
- Greater community involvement to reduce vulnerability (e.g. Neighbourhood Watch systems in UK to look out for elderly in cold spells).

Political adjustments

- Emphasising evacuation plans, services and emergency centres (e.g. these were increased in the Gulf of Mexico following Katrina)
- Land use zoning and restrictions (e.g. along the Mississippi in New Orleans following Katrina)
- Issuing early warnings and coordinating emergency services (e.g. hurricane preparation in south-east USA)
- Spreading economic loss via insurance, grants, etc (e.g. US government in New Orleans following Katrina).

Remember

Many so-called primitive cultures have lived with climatic hazards for a long time. They have adjusted their ways of life to the hazards, thereby minimising the costs.

Managing climatic hazards effectively depends upon:

- The nature of the hazard – its severity, scale, frequency, any build-up signs, etc
- The level of preparation and awareness of risks
- The nature of the area – its structure, geology, relief, climate, etc
- The level of development – research, technology available, communications, warnings, etc
- The nature of the population – density, education, mobility, level of perception, etc
- Political organisation – coordination, priority, existence of emergency plans, etc.

Case study: drought management in Chad

Student book pages 111–12

Causes	◆ High pressure Sahel area ◆ Inland ◆ Desertification – bush felling and overgrazing with goats ◆ Commercialisation of farming (e.g. cotton, which needs water for irrigation)
Development and management options	
Water development	◆ Water transfers but high cost and huge water loss ◆ New dams – cost and where? ◆ Develop groundwater sources – limited and will lower water table ◆ Re-use effluent and recycle grey water

Case study: drought management in Chad *(continued)*

Management	◆ Reduce water losses – cover aqueducts, etc ◆ Increase cost of water ◆ Ration water ◆ Reduce water use (e.g. dryland farming, drought-resistant animals, etc) ◆ Better technology (e.g. irrigation of individual plants) ◆ Low level technology (e.g. walls of stones to slow runoff)

Quick check questions

1 Why does afforestation reduce climatic hazard impacts?

2 Why is hurricane seeding to reduce hurricane strength not more widespread?

3 Why are laws needed to enforce evacuation of areas under threat?

4 What was the main reason Hurricane Katrina did so much damage in New Orleans?

5 Why do people live in areas that are prone to hurricanes?

6 Why are governments reluctant to give early warnings of a possible hurricane?

7 What is the chief cause of the drought in Chad?

Remember

You might like to compare Chad's approach to drought to the UK's approach.

3.5 In what ways do human activities create climatic hazards?

Student book pages 113–19

Key words

Global warming
Greenhouse gases
Long-wave radiation
Global dimming
Particulates

Global warming

Greenhouse gases in the atmosphere trap the Earth's long-wave radiation. Earth's temperature is predicted to rise by 1–6°C this century.

Causes

- Burning fossil fuels produces carbon dioxide (which will have doubled between 1950 and 2050)
- Transport producing carbon dioxide and nitrous oxide
- Farming – cattle, rice paddy producing methane (a major greenhouse gas) and agrichemicals (e.g. fertilisers) releasing greenhouse gases
- Deforestation – less carbon absorbed and often cleared by burning
- Waste tips produce methane
- Industry producing hydrofluorocarbons (increasing at 4 per cent a year)
- Melting permafrost releasing methane and carbon dioxide.

Once underway there are a number of feedback mechanisms that reinforce the process (e.g. melting of ice further reduces the reflectivity of the Earth's surface).

Effects

- Sea level rise leads to flooding of low-lying coastal areas and increased coastal erosion
- Increased flooding (coastal and river)
- Increase in extreme weather as more energy in atmosphere, leading to increased storms and hurricanes

- Change in crop patterns, leading to increased risk of famine
- Reduced rainfall in some areas, leading to spread of deserts and droughts
- Change in natural vegetation (e.g. pine trees grow further north)
- Change in economic activity (e.g. fewer winter sports resorts)
- Wildlife will adapt or become extinct; pests and diseases will spread (e.g. malaria into Europe).

Exam tips

You will probably need to draw a diagram to show the differing impacts the greenhouse gases have on in-coming short-wave radiation and out-going long-wave radiation.

Remember

Not all the effects are negative for all communities. Some areas will gain from global warming in the short term.

Remember

Global warming is different from the destruction of the ozone layer. Many candidates confuse these terms.

Global dimming

In the 1980s there was 10 per cent less sunlight than in the 1960s but it is now rising.

Causes

- Heavy industry producing particulates means that more clouds form
- Heavy industry producing sulphate pollution that reflects sunlight
- Could be a side-effect of global warming.

Effects

- May offset global warming
- Reduced effectiveness of solar power
- Slower growth of crops and lower yields
- Human health problems – Seasonal Affective Disorder (SAD) and lack of vitamin D.

Case study: acid rain in China

Student book page 117

Background	◆ 2003 – caused US$13 billion losses ◆ 33% of China suffers acid rain ◆ Worst in south, with rain with a pH of 5
Causes	◆ Increased car ownership ◆ High usage of coal, as China lacks other fuels – 75% of power stations are coal fired ◆ Rapid industrialisation ◆ Over-use of nitrate fertilisers ◆ Rapid urbanisation = higher demand for power ◆ Lax anti-pollution laws and little environmental protection
Impacts	◆ Increased chemical weathering – buildings, roads, etc ◆ Rail and power lines corroded and fail ◆ Soils increase acidity = reduced crop yields ◆ Loss of forests = increased soil erosion ◆ Rivers and lakes increase acidity = damage to wildlife and fish farming ◆ Eye and respiratory problems increase ◆ Ancient buildings corrode
Solutions	◆ Government taxes pollution and issues licences to discharge pollutants ◆ New technology to allow clean coal burn and desulphurisation ◆ Reducing fossil fuel use, e.g. Three Gorges HEP scheme ◆ Adding powdered limestone to lakes ◆ 2006 – US$175 billion invested to reduce pollution levels

Key words

Dry deposition

Wet deposition

pH

Remember

If you wish to investigate micro-climates as part of your practical work for the skills paper, this is the area of the specification it comes under.

Case study: photochemical smog, Los Angeles

Background	◆ Los Angeles was the second most polluted US city in 2008 ◆ The photochemical smog resulted from nitrous oxide reacting with hydrocarbons in intense sunlight
Causes	◆ 11 million vehicles (24 million in state of California) ◆ Sunny dry climate – only 35 days a year are wet (rain removes pollution) ◆ Temperature inversions trap the fumes ◆ Basin effect of the relief causes pollution to fill it ◆ Ports (e.g. Long Beach) = diesel fumes ◆ Lack of public transport ◆ Roads align with prevailing wind
Impacts	◆ 1600 premature deaths (due to respiratory disorders) each year ◆ Children have 10–15% reduction in lung capacity ◆ Absenteeism from work – loss of workers ◆ In 1970s there were dangerous levels of smog on 100 days per annum ◆ Crops wilted in fields due to lack of light ◆ People suffered from depression ◆ Forced Hollywood film-makers to make films elsewhere
Solutions	◆ By 2001 there were no days recorded as dangerous to health because of legislation and regulation ◆ Clean Air Act 1963 – reduced smoke pollution ◆ 2004 – ban on all outdoor burning ◆ 1966 – exhaust controls on cars ◆ California Air Resources Board (CARB) set up in 1967 and set air quality standards in 1969 ◆ CARB aims to reduce petrol use by 20% in next 10 years (biofuels and electric vehicles) ◆ 1990s set the strictest standards for low emission vehicles – 10% of vehicles were zero emission by 2003 ◆ Regular smog checks carried out, with fines if vehicles didn't meet standards

ExamCafé

Student book pages 84–123

Section A

Sample question

Study resource 3, a diagram comparing the relative costs of droughts in Kenya and Australia, showing higher human costs in Kenya and higher economic costs in Australia.

*Identify **one** issue and suggest appropriate management strategies to deal with it*
[10 marks]

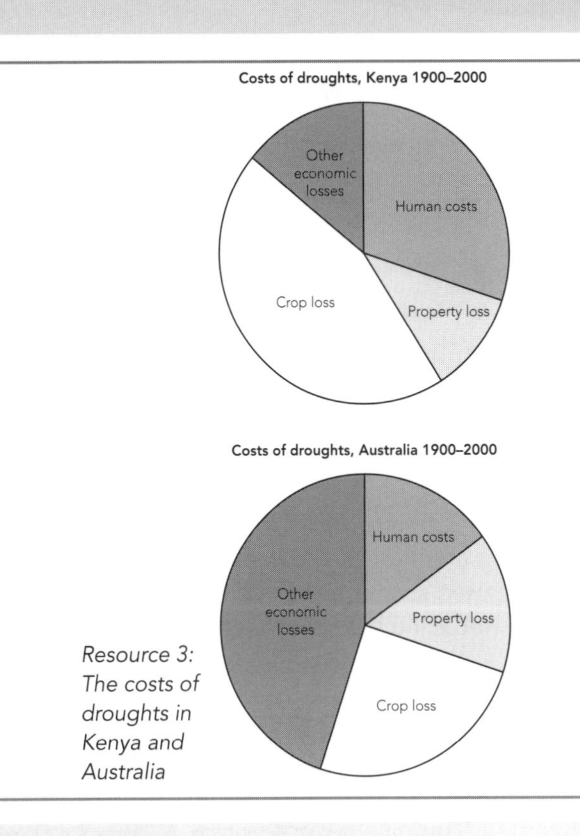

Resource 3: The costs of droughts in Kenya and Australia

Student Answer

The issue is the relatively higher economic costs to human costs in Australia compared to Kenya.

This is fine as far as it goes but the answer needed to refer more to the resource and quote some figures to give some idea of the scale of the contrast.

When it comes to management strategies, you need to make a choice – is it about closing the 'cost gap', or solving drought or making areas more capable of coping with drought? Clearly this reflects back on how you express the problem so in hindsight a better answer might be:

The issue is that droughts tend to cause a higher human cost in a LEDC like Kenya compared to a MEDC such as Australia.

Now the management focus is clearly on strategies to reduce the high human cost in LEDCs:

A number of strategies can be used to reduce the high level of human costs. Some are aimed at reducing the severity of the drought such as greater water storage in wetter periods, water transfers from wetter areas, use of underground water, etc, whilst others seek to adapt the population to drought conditions. The latter strategies are probably more effective in the long term and can be of a low-level technology/cost type. These could include changing farming practices to more drought-resistant versions such as replacing cattle with goats, use of mulch to keep water in the soil, etc.

This answer has many strong points and does seek to evaluate the two types of strategies but:

- It lacks details of the strategies
- It lacks examples
- Are these examples of low-level technology?

Section B

Sample question

Evaluate the effectiveness of strategies used to reduce the impact of tropical storms. *[30 marks]*

Like all evaluation questions, this one implies that some of the strategies are more effective than others. You should always start by reading the question carefully. There are a number of key words here:

- 'effectiveness' – in what sense and to whom?
- 'strategies' – more than one type is expected, as the term is plural
- 'reduce' – not stop but rather mitigate
- 'impact' – can refer to primary or secondary impacts or economic versus social versus environmental
- 'tropical storms' – really means hurricanes and cyclones so don't get too 'picky' over terminology.

This clearly draws on two sections of the specification – section 4 for strategies but also section 1, as it emphasises the impact of tropical storms. To some extent this is a straightforward list of strategies used such as prediction, risk assessment, etc, but the key thing is to evaluate their effectiveness in reducing the severity of the storm's impacts. For each of the strategies it is therefore crucial to present evidence as to when or where they have been effective and where they haven't:

Student Answer

The use of prediction is often very effective. Tropical storms are very visible. The use of weather satellites and increasingly sophisticated models of storm movements often give three to four days warning of a storm as in the case of Hurricane Katrina for New Orleans in August 2005. This gives more time in which the area can prepare and so reduce the impacts. Unfortunately prediction isn't always accurate, especially over the exact timing and location of impact. Hurricane Ivan in 2004 was not predicted to double back on itself. More crucial is whether anything results from the prediction. In the case of the three-day warning for Katrina the Governor of New Orleans waited until 19 hours before the predicted impact before ordering a mandatory evacuation of the city.

This is an effective paragraph, with a clear evaluative focus and clear, if a little thin, exemplification. What lifts it to a high level is the point that prediction means little unless someone does something about it. The candidate could have made a further point that prediction helps people to prepare and so reduce impacts on their activities but does little to help non-human aspects or fixed items such as bridges, power lines, etc.

More effective answers always demonstrate a clear structure. Here, an effective one might be to group strategies into:

- Before the event, e.g. forecasting
- During the event, e.g. storm shelter management
- Immediately after the event, e.g. search and rescue
- Long term, e.g. raising sea walls, planning, etc.

Quality answers may seek to explain why the effectiveness of strategies varies over time and between locations. Many will adopt a LEDC versus MEDC approach but there are other factors to consider, especially the size and ferocity of the storm.

Chapter 4
Population and resources

Student book pages 124–61

4.1 How and why does the number and rate of growth of population vary over time and space?

Student book pages 126–32

Exam tips

Population is a system with inputs of births and in-migration and outputs of deaths and out-migration. The balance of these inputs and outputs changes the nature and structure of the population.

Population growth is: natural increase plus net migration.

Natural increase is: birth rate (BR) minus death rate (DR).

Net migration is: in-migration minus out-migration (the result of push versus pull factors).

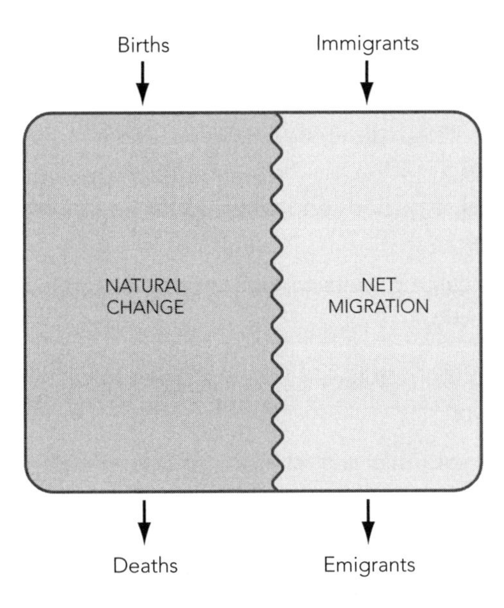

The input–output model of population change

Factors influencing population growth – natural increase

- **Physical** – hazards can increase death rates, diseases and climate influence BR and DR
- **Demographic** – age structure (e.g. young population = high BR), sex ratio, infant DR, ethnicity
- **Social** – health, socio-economic status, culture, status of women, religion, education and literacy, level of overcrowding, social services, age of marriage, marital status, alternative attractions (materialism)
- **Economic** – income, employment type, infrastructure (e.g. transport), standard of living, cost of children, income from children, nutrition, housing quality
- **Political** – policies (e.g. birth control), security (war), laws (e.g. women's status), tax policy (e.g. family allowance), indirect (e.g. education), old age pensions, inheritance

Remember

These factors influence both death rates and birth rates. Always look at the demographic factors first, as these are the most important influencing factors.

Remember

Birth and death rate figures are often crude – simply giving the number per thousand population in a year. They are crude because they ignore the sex and age pattern of the population.

Population changes over time – the Demographic Transition Model

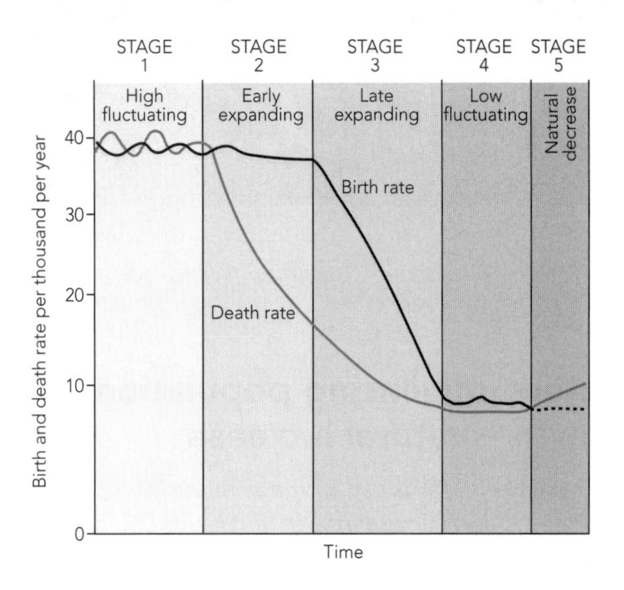

This model is based on the concept that death rates and birth rates are linked. If the death rate starts to fall (due to better diet, more healthcare, etc) then there is less need for a high birth rate to offset deaths (especially infant deaths). There is always a time lag, as people need to be sure it is a long-term decline in the death rate.

Changes over time in natural increase	
High fluctuating	High BR/DR, DR fluctuates due to wars, famines, etc; little growth, life expectancy low, rural population (e.g. Amazon Basin)
Early expanding	DR declines due to better nutrition, public health, etc; BR high, rapid growth, life expectancy rising, urbanisation starts (e.g. Bangladesh)
Late expanding	BR falling, DR low, BR falls as life expectancy increases and lower infant DR, rapid growth, rapid urbanisation (e.g. Brazil)
Low fluctuating or stationary	Low BR/DR – fluctuates with economic growth, slow if any growth, counter-urbanisation (e.g. Taiwan)
Natural decrease	BR now below DR – BR low as ageing population, materialism, working women, etc, declining population, counter-urbanisation (e.g. Hungary)
BUT	◆ Limitations of model – as applied to LEDCs today, as based on Western Europe's historical process ◆ DR has fallen more rapidly, as advances in technology are transferred and applied immediately ◆ Less clear cause/effect link to economic development ◆ Many countries now starting from higher BR than in the past ◆ Very different cultural, social and political conditions apply today ◆ Link between DR and BR can be challenged

Key words

Natural increase
Net migration
Demographic transition model (DTM)

Factors influencing population growth – net migration

Forced

Migrants have no choice in or control over migrating, e.g. expulsion of Ugandan Asians

Voluntary

Pushed out by perceived disadvantages of country of origin:

- Demographic, e.g. lack of marriage partners
- Physical, e.g. famine, drought, natural disaster
- Economic, e.g. unemployment, poverty, poor infrastructure
- Social, e.g. poor education, poor health, violence, crime
- Political, e.g. persecution, insecurity, limits on freedom.

Pulled to the perceived advantages of the destination:

- Demographic, e.g. availability of marriage partners
- Physical, e.g. better climate, fertile soils, attractive scenery
- Economic, e.g. jobs, higher wages, better housing, lower taxes

- Social, e.g. good services, safety and security
- Political, e.g. democracy, freedom, security.

But there are pressures that hold migrants back:

- Transport availability – build more roads and more can migrate
- Cost – it costs a lot to migrate (very poor are inert)
- Knowledge – migrants need knowledge of possible destinations (via media, friends, relatives)
- Inertia, e.g. those with relatives in an area are reluctant to leave them.

The decision to migrate is therefore often influenced by:

- Age – young are more mobile (fewer responsibilities)
- Sex – traditionally, males are more mobile but in many MEDCs it is females who are more mobile, reflecting job opportunities
- Family status – childless are more mobile
- Education – higher educated are more mobile (brain drain)
- Wealth – poorest can't afford to migrate.

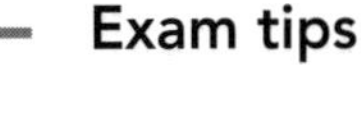

Exam tips

As birth and death rates are falling all over the world, net migration is becoming the dominant factor influencing the number and structure of the population in a country or area.

Key words

Emigration
Immigration
Intra-urban
Commuting
Circulation

Changes over time in migration – Zelinsky's model, 1971 (links to DTM)		
Phase	Name	Migration
1	Pre-modern traditional society – agricultural	Limited migration and circulation
2	Early transitional society – industrialisation	High population growth so high rural to urban, colonisation of frontier lands, emigration
3	Late transitional society – growth of tertiary sector	Population growth slows so migration slows but circulation (e.g. commuting) increases and immigration may take place
4	Advanced society – decline in heavy industry	Inter-urban and intra-urban dominate, increasing circulation
5	Future advanced society	General decline in migration but urban to rural more common and high circulation (often international) may increase, e.g. tourism and commuting

Migration into the UK since 1 May 2004

On 1 May 2004, eight mainly eastern European countries joined the EU and their inhabitants gained freedom of movement within the EU. Estimates vary but some 500 000 immigrants are thought to have arrived in the UK, with the largest number (63 per cent) coming from Poland. A survey showed that the vast majority of migrants came for work and to access higher standards of living than in their home countries. A Treasury report in 2007 calculated they had added a net £6 billion to the UK economy.

What impact have these migrants had on the UK?

- 82 per cent were aged 18–34 so they reversed the ageing population
- Reduced the dependency ratio by 8 per cent
- The male/female ratio was 58/42 so they increased the number of males, leading to intermarriage
- 25 per cent of births are to recent immigrants (75 per cent in some areas)
- Made the economy more flexible and responsive (20 per cent had temporary agricultural jobs)
- Filled jobs that the local population didn't want (26 per cent worked in distribution, hotels and catering)
- Increased demand for cheap housing in certain areas – migrants tend to cluster
- Increased need for schools
- New types of shops, cafés, etc, opened – cultural injection

- Generated revenue – over 10 per cent of government receipts (tax)
- Minimal friction as they share European culture, e.g. increased church attendance.

The Treasury report suggested in 2007 that 17 per cent of the UK's economic growth since 2004 was due to immigration.

Exam tips

In an essay in this section your main aim should be to evaluate how and why population changes in response to a number of factors or to evaluate the relative importance of those factors.

Concepts of overpopulation and underpopulation

These are relative terms because an area is overpopulated relative to something – usually some aspect of resources, e.g. food supply.

Overpopulation is: when population exceeds resource and technology levels so the standard of living falls

Optimum population is: the level of population that produces the highest standard of living

Underpopulation is: when there is insufficient population to use the resources of an area efficiently

Population pressure is: when the population exceeds the area's carrying capacity

Remember

These questions are largely evaluative so you may have to evaluate the notion that too many people cause overpopulation. This is a Malthusian viewpoint but there are alternative views. Also remember that a country can only be overpopulated relative to something.

Population theories

	Malthus	Boserup
Date	1798	1965
Viewpoint	Deterministic	Possibilistic
Trend	Population will grow until it exceeds resources (food)	Population will grow until it puts pressure on existing resources
Then	Positive checks operate that increase DR (e.g. famine, war, disease) so population decreases	Once there is pressure on resources, innovation increases production so population can continue to grow
Reaction	Preventative checks to slow population growth, e.g. delayed marriage	Through innovation or intensification, humans can increase their resource base
Evidence	Resources have increased more rapidly than Malthus predicted so populations have grown but in some areas famines support his view	Resources have increased more rapidly than Malthus predicted but today there is increasing pressure on finite resources at a global level

Key words

Optimum population
Possibilistic
Malthusian
Deterministic
Population pressure
Newly Industrialised Countries (NICs)

Classifying countries

It is possible to classify countries on the basis of population and resources. Here are some examples.

Resources	High population		Low population	
	Rapid growth	Slow growth	Rapid growth	Slow growth
Abundant	Brazil, Nigeria	Russia	Congo	Australia
Scarce	Bangladesh	UK	Yemen	Mali

Global contrasts in population growth and policies

	LEDC – Bangladesh	NIC – China	MEDC – UK
Population	140 million	1332 million	61 million
Birth rate	25/1000	13/1000	4.9/1000
Death rate	8.5/1000	7/1000	6.9/1000
Fertility	2.72	1.7	1.9
Infant mortality	68/1000	20/1000	5/1000
Life expectancy	60	73	79
Net migration	– 0.75/1000	– 0.41/1000	+ 0.20/1000
Population growth	+ 1.6%	+ 0.5%	+ 0.6%

Some countries take action to control the rate of population growth for economic reasons (e.g. to increase the supply of labour), social reasons (e.g. to reduce the demand on education and health services) or political reasons (e.g. to have a larger army).

Policy	Action	Example
Antinatalist	Free family planning, limits on family size (taxing larger families), education, advertising	Iran
Pronatalist	Offering tax incentives, grants and rewards for large families	France
Family friendly	Using childcare, maternity and paternity leave from work to give parents a choice	UK
Limits on migration	Using quotas and work permits, must have relatives in the country	USA
Selecting migrants	Only allowing in those with required skills, income, age, etc	Australia

Key words

Antinatalist
Pronatalist
Quotas

China – an example of non-democratic state intervention

- State encouraged high BR in 1950s as 'large population gives a strong nation'.
- State family planning programmes introduced in 1970s – 'later, longer, fewer' – later marriages, longer gaps between children. Needed to reduce unemployment and use of resources, e.g. an estimated 20 million died in famines in the 1960s.
- 1979 – one child per family policy. Inducements were free education, priority housing, pensions, etc. Disincentives were loss of inducements and fines of 15 per cent of income. Marriage age raised to 20 for women and had to apply to the state for permission to have a child.
- 1980 – abortion compulsory for second pregnancy. State-run family planning and close monitoring.
- 1987 – government slightly relaxed the rigid policy. People in rural areas could have a second child if the first was a girl or disabled.

By 2008 China had reached 1332 million, with a static birth rate of 13/1000 and a low death rate of 7/1000. The annual population growth rate is 0.55 per cent.

Impact of one child per family policy

- 300–400 million fewer people
- Sex imbalance, as Chinese culture favours boys (117 males per 100 females)
- Abortions – 10 million a year (98 per cent female)
- Adoptions abroad (e.g. USA) – chiefly girls
- Increased savings for old age as no support from children for elderly
- More female workers
- Increased dependency ratio – ageing population
- 'Little emperor' problem – spoilt single boys
- Increased rural-urban contrast
- Human rights issues.

Quick check questions

1 Why does increasing the rights and education of women reduce the birth rate?
2 Why might the DTM not be followed in many LEDCs today?
3 Why might improved accessibility a) reduce the population b) increase the population of an area?
4 What do you think is the link between Zelinsky's model and the DTM?
5 Why might increased immigration cause a government political problems?
6 How did Australia initially reduce its chronic level of underpopulation?
7 Which part of the world might you quote as an example of Malthus' ideas being valid today?
8 What was the chief problem caused by China's one child policy?

4.2 How can resources be defined and classified?

Student book pages 133–7

A **resource** is: any aspect of the environment that can be used to meet human needs.

Classifying resources

Resources are classified by:

- **Origin** – human versus natural, animal versus vegetable/mineral
- **Source** – organic versus inorganic, extraterrestrial versus terrestrial, surface versus sub-surface, domestic versus imports, point versus diffuse
- **Form** – liquid, solid, gaseous, visual, etc
- **Use** – high value versus low value, stores versus flows, vital versus non-vital
- **Ubiquity** – scarce versus ubiquitous (e.g. gold versus aluminium)
- **Extent of renewability** – renewable (e.g. water), non-renewable (e.g. fossil fuels), semi-renewable (e.g. fish)
- **Ownership** – private versus state, notion of belonging to all (e.g. fish).

But many of these classifications are interconnected.

Changing resources

Resources change as:

- Technology changes (e.g. uranium wasn't a resource in the 18th century)
- Ability to exploit develops (e.g. offshore oil fields)
- Population grows in size (e.g. land)
- Substitution – replacement of less efficient resources (e.g. water power by steam)
- Price or cost changes (e.g. increase in oil price)
- Education increases/attitudes change (e.g. scenery now seen as a resource)
- Restrictions put on importing or exploitation of resources (e.g. artificial rubber made from dandelions during Second World War).

Key words

Resource depletion
Technology
Resource management

The definition of resources may change due to:

Changes in the technology used to

- Find the resource
- Extract the resource
- Transport the resource
- Process the resource
- Use the resource
- Waste-manage the resource
- Recycle or re-use the resource.

Social changes including

- Increased population size
- Raised education levels
- Changes in culture
- Changes in attitude to pollution, waste, etc
- Changes in expectations and tastes.

Key words

Ubiquitous
Semi-renewable
Natural resources

4.3 What factors affect the supply and use of resources?

Student book pages 137–43

Defining supply and use

Remember

'Supply' and 'use' are not exact terms but carry a number of meanings in terms of resources. This is a useful point to recall when asked to evaluate some aspect of supply and/or use.

What is meant by supply?

- A resource that exists in the ground or ocean – some finite (e.g. oil), some variable (e.g. crops)
- The amount extracted
- The amount available for use (e.g. metal ore content, level of waste in crops)
- The amount available to consumers.

What is meant by use?

- Used directly by consumers (people, industry, etc)
- Used indirectly as it produces by-products or forms only one stage in producing the final product, e.g. timber.

Some factors that influence supply	
1. Physical conditions of an area or the resource	
	Climate – type, extremes
	Geology – rock and mineral type, mineral content, structure, faults, hardness
	Water – quantity and quality, seasonal variation
	Relief – flat versus steep, roughness of the terrain
	Soil – type, depth, fertility, drainage
	Biotics – plants, animals, etc
	Distance – remoteness
	Accessibility – how easy it is to reach or transport the supply
2. Nature of the resource	
	Scale and size of resource (quantity and quality)
Often these overlap, e.g.	Granite = poor, wet, acid soils = moorland Chalk = thin, well-drained alkaline soils = grassland
3. Human factors encouraging resource development	
	Technology – equipment, processes, detection, sustainability
	Capital – quantity, type (e.g. fixed versus operating, long term versus short term)
	Industry – type, demand for raw materials, investment and ownership
	Transport – type, bulk carrying capacity, cost, direction, time
	Demography – workforce, technical know-how
	Energy – type, volume, cost, consistency
	Alternatives – competition, by-products, role of imports, recycling
Others	Conservation and protection, cost of making good, pollution
	Social or cultural attitudes – religious views, tradition, inertia
Often these overlap with the physical, e.g.	Chalk = thin, well-drained alkaline soils = grassland = sheep = if fertilised then ideal for arable crops

Exam tips

It is rare for these factors not to overlap. It makes economic and practical sense to exploit those resources first that offer the greatest advantages, in an area where they are most accessible.

How and why do these factors change over time?

Physical factors are often constant (and finite) but are subject to short-term fluctuations such as droughts, which in turn impact on other physical resources (e.g. biotic). Long-term climatic change is having a major impact on the supply of resources (e.g. growing numbers of vineyards in England). Also, many resource supplies are becoming exhausted, as finite supplies are used up (e.g. copper).

How long will resources last?

The following table shows how long a selection of resources will last if they continue to be consumed at the current rates.

Mineral	Use	Main known reserves	Years left (in 2007)
Antimony	Drugs	China: 62%	13
Chromium	Chrome plating	South Africa: 35%	40
Copper	Wiring, coins, plumbing, electrics	Chile: 38%	38
Lead	Pipes, batteries, fuel additive	China: 25%	8
Tin	Cans, solder, electronics	China: 31%	17

Source: New Scientist, 26 May 2007, pages 34–41

Remember

When resources are almost exhausted their price tends to soar. This encourages the search for new sources, substitutes or reclamation of the original resources.

Human factors change more rapidly, often as a direct result of demand changes or limitations in supply (very much as suggested by Boserup and the resource optimists). For example, chalk uplands were once largely used for sheep farming but then the introduction of manure and fertilisers allowed them to be used for wheat farming. Similarly, four-course crop rotation and the use of marling transformed the poor sandy heathlands of north Norfolk into major arable farming areas in the 19th century.

Reasons why the factors controlling resource supply change

The table opposite shows the reasons behind changes in the factors controlling supply of a selection of resources.

Exam tips

You are expected to have studied two different resources. One of these can be an energy resource so you may wish to look back at some of your AS work on oil or coal, for example.

Reason	Example
New technology invented	The rise of micro-electronics made silicon a valuable resource
Existing resource more difficult to get	Aluminium replaced scarcer copper in cables
Cheaper foreign imports	Decline in cattle numbers due to cheap dairy products imported into the UK
Increased mobility of population	The appreciation that natural scenery is a resource
New areas of resources discovered	Offshore oil fields found (e.g. in the North Sea)
New processes developed	Use of recycled waste paper to produce paper
Increased education of population	Inventiveness of the population seen as a resource (e.g. in Switzerland)

Case studies: resource supply in western USA and Cornwall

	Water	Tin and copper
Location	Western USA **Student book pages 138–42**	Cornwall
Factors	Covers 60% of area of USA, has 40% of US population but receives only 25% of country's rain Demand greater than supply	1840–60: largest supplier of tin and copper in the world Supply exhaustion
Physical (supply)	◆ Low rainfall, as continental or Mediterranean climate ◆ Series of droughts and high evaporation from lakes behind dams ◆ Excessive pumping from aquifers and groundwater ◆ Forestry has reduced evapo-transpiration so drier	◆ Vertical veins in granite – numerous but scattered ◆ Ore metal content quite low (8%) so produced a lot of waste ◆ Veins went deep and even out under the sea so needed pumping ◆ Hard tough ore to mine ◆ Found with valuable trace minerals, e.g. arsenic
Human (use)	◆ Rapid population growth especially city development, e.g. Los Angeles ◆ Increased affluence – golf courses and swimming pools use water ◆ Expansion of irrigated farming – 80% of the water ◆ Large concentrations of water-using industry, e.g. steel, chemicals ◆ 25% of water moved is lost in leakages	◆ Supplied much of the 19th-century British industrialisation ◆ Discovery of electricity made copper very valuable ◆ High value compared to weight of ore so later transported to coalfields for smelting ◆ By 1862, 340 mines employed 50 000 people ◆ Lack of local energy supplies to power pumps ◆ Cheaper deposits more easily extracted in Malaysia in 1870s

Case studies: resource supply in western USA and Cornwall (*continued*)

Outcomes	◆ Dams on the Colorado ◆ Water transfers to west via canals and aqueducts ◆ Development of more groundwater resources ◆ Weather modification, e.g. cloud seeding ◆ New sources of surface water, e.g. desalinisation, icebergs ◆ Recycling of water, e.g. sewage water for golf courses ◆ More efficient use of water, e.g. short-flush toilets ◆ Charging more for water, e.g. farmers only pay 10% of the cost ◆ Improving efficiency of irrigation, e.g. drip irrigation ◆ Changing from water-dependent crops to ones needing less water	◆ Mines closed – all gone by 2005 ◆ Re-working of spoil heaps for rare minerals ◆ Some mines continue as tourist attractions, e.g. Poldark mine ◆ Once price has risen, move to re-open South Crofty mine ◆ Mining towns such as Cambourne suffered unemployment and out-migration ◆ Visual pollution of waste heaps and old engine sheds ◆ Danger of hidden mine shafts ◆ Did leave higher than expected level of transport links

Quick check questions

1 Why are mobile resources such as fish or water so problematic?

2 Why do you think that war has often been referred to as the 'mother of invention'?

3 a) Why is hemp (a crop) no longer grown to make sacks and rope?
b) Why might it be grown again in the future?

4 Some resources (e.g. oil) are very unevenly distributed in the world. Why might this lead to problems?

5 Can you suggest why we use so much aluminium?

6 The closed system of the water cycle ensures that there is a constant supply so why are there shortages?

7 Why are tin mines in Cornwall being re-opened?

4.4 Why does the demand for resources vary with time and location?

Student book pages 144–48

Demand is: the amount or volume consumers want and it also includes the quality. Demand tends to be reflected in the price and availability of a resource.

Remember

Price is the mechanism that reflects demand in capitalist countries but countries with a different culture can and do use different indicators of demand, e.g. the state's own priorities.

Remember

Demand depends on both quality and quantity. To some extent, these aspects can be substituted for one another, so less of a high-quality resource may equate to more of a low-quality resource.

Exam tips

Be careful when you choose two resources to compare in an examination answer. If they are too dissimilar (e.g. fish and gold) it may be more difficult to compare them.

Factors affecting the level of demand for a resource

Factor	Example
Its price or cost	Petrol price rise has reduced levels of demand
Its value	Gold is greatly in demand, as it is rare and valuable
Its importance	Oil is considered vitally important for energy production
Its availability	Water is ubiquitous in the UK so demand is relatively low
Its purity or content	Pure, clean water is needed by the brewing industry
Its ease of use	Coal is demanded as a power station fuel by LEDCs, rather than the more difficult to use uranium
Its ease of transport	Oil and gas are easier to transport than coal
Its level of waste	There is a decreasing demand for dirty, polluting coal

Key words

Production
Extraction
Pollution
Marketing

Demand also varies with human factors:

Factor (human)	Example
Its importance	Modern societies rely on petrol for travel and transport
Its reputation	Gold is in great demand, as it is rare and valuable
Its uses	Oil can be used to make a range of chemical products
People's perceptions	Platinum is an attractive metal for jewellery
Sheer number needing it	Food supplies, especially basics like wheat and rice
Tradition	The British drink a lot of tea
Marketing	Special offers etc increase demand
Advertising	Advertising in the media encourages increased demand

Case study: a comparison of two minerals in the Peak District, UK

	Limestone	Fluorspar
Use	56% road chippings, 23% cement and 17% chemicals	35% in steel industry, 30% chemicals, e.g. solvents, toothpaste, anaesthetics
Production	6 million tonnes a year	60 000 tonnes a year
Characteristics	◆ Heavy, bulky ◆ Low value ◆ Common rock in UK ◆ Can use alternatives, e.g. chalk for cement ◆ Easy to use – mechanical crushing	◆ Light, fine ◆ High value ◆ Rare mineral in UK ◆ No alternatives, economic ◆ Complex chemical processing needed
Extraction	◆ Quarrying – large surface impact ◆ Explosives and giant machinery needed ◆ Large waste heaps	◆ Mined – little surface impact ◆ Explosives but little surface machinery ◆ Little waste, as used to fill mines
Transport	Large lorries	Lorries but fewer needed
Environmental impact	Large – produces dust and noise, carves away hills	Little, as small scale and underground

Demand varies over time, as:

- Population grows – a larger population needs more resources
- Population structure changes – e.g. an ageing population has different demands, increase in ethnic migrants, etc
- Areas develop – demand structure changes, e.g. mechanisation of farming – employment sectors change over time
- Wealth increases – as incomes rise, people demand more consumer goods, travel more, consume higher-quality, more varied food, etc but this varies with their income elasticities
- Globalisation – ascent of Rising Industrialised Countries (RICs) and NICs has put pressure on resources
- Governments change – e.g. left-wing versus right-wing views, communism versus democracy, totalitarian government = different demands

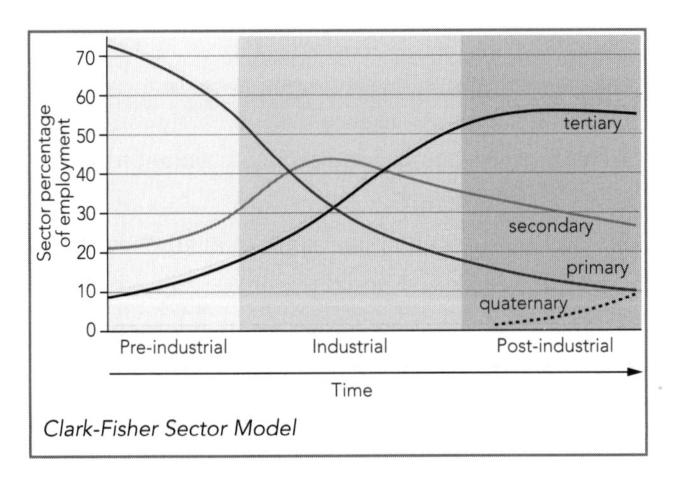

Clark-Fisher Sector Model

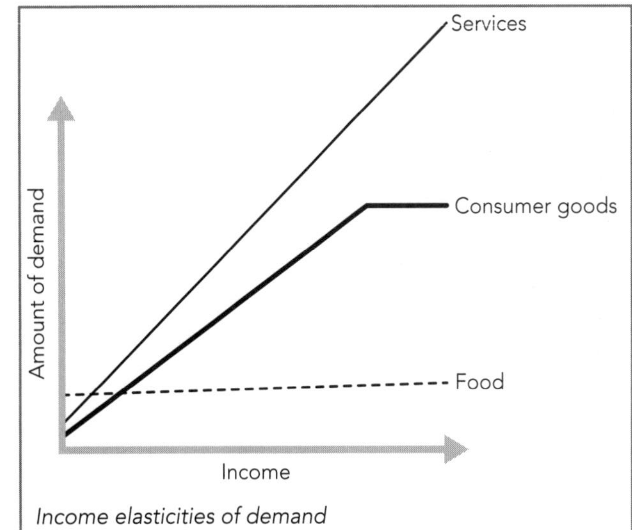

Income elasticities of demand

Demand also varies according to location, with:

The nature of the resource	
Key factor	Cost/price/value
Quality of the resource	Purity, waste content, is it scattered?
Quantity of the resource	Volume, weight, life expectancy
By-products	Does it come with valuable by-products, e.g. cotton comes with fibre and an oil seed
Importance	Is it vital, how dependent are we on it (e.g. oil)?
Location	Is it near or remote (e.g. Alaskan oil)?
Substitutes	Are there possible substitutes (e.g. aluminium can replace copper)?
Flexibility	How many uses does it have?
Ease of transport	Bulk, perishability, hazardousness, cost
The nature of the area	
Technology	Ease of extraction, processing, etc
Conditions	Do the physical conditions aid or hinder extraction? Or do they require certain resources (e.g. energy in cold areas)?
Demography	Is there a local labour force?
Standard of living	Basics or necessities are demanded first
Culture and tradition	May conflict with extraction
Accessibility	Core versus periphery
Impact on area	Need for protection or conservation
The socio-political climate	
Government policy – planning, nationalisation, etc	
Security and safety	
Social attitudes	

Contrasting patterns of demand

	LEDC	NIC	MEDC
	Bangladesh	China	USA
Population	150 million	1300 million	301 million
Population growth (per cent)	1.6	0.5	0.6
Oil consumption (barrels per day per 1000 people)	0.5	5	69
Meat consumption (kg per person)	3	52	125
Average water use (litres per person per day)	46	86	576
Annual electricity consumption (KWh per person)	120	2140	12 200

Remember

Comparisons can be very misleading. Not all cultures value resources or products in the same way. For example, people who follow certain religions will not eat pork so their meat consumption might appear lower than expected.

Ecological footprints

Student book page 145

The impact of an individual (or a country or region) on resources is often referred to as the ecological footprint – the equivalent area of land required to produce the resources needed per person. For example, someone living in North America has a footprint equivalent to 9.5 ha and someone living in Africa has a footprint of 1 ha.

Exam tips

'Ecological footprint' is not in the specification so you cannot be asked a question directly on it but it is a useful concept to introduce into an evaluation of resource demand or supply/use.

Demand is very uneven across the world. In an average lifetime:

	An American consumes	A Malian consumes (as a percentage of American consumption)
Phosphorous	8322 tonnes	1%
Copper	630 kg	5%
Aluminium	1576 tonnes	7%
Tin	15 kg	3%
Gold	48 g	12%

So what would happen if everyone adopted the American lifestyle?

Quick check questions

1. Why is an ageing population a problem for medical resources?
2. Why is tertiary employment not at zero level in a pre-industrial society?
3. Give an example of a resource that is inelastic in demand (doesn't change in demand as its price rises or falls).
4. Why might the demand for steel rise as incomes rise?
5. Why do LEDCs consume much less meat than MEDCs?
6. Why do MEDCs consume so much water? (A person in the USA consumes over 12 times as much as someone in Bangladesh.)
7. How much greater is the ecological footprint of an American than that of an African?

4.5 In what ways does human activity attempt to manage the demand and supply of resources and their development?

Student book pages 149–56

Management:

- May be governmental, individual or corporate, e.g. TNCs
- Free markets find such controls difficult but totalitarian states have few problems.

Why bother?

- Help control prices, for health or security, avoid exhaustion of resource, reduce overdependence, protect trade balance.

Possible methods

1. Control demand:

- Rationing, quotas – on basis of need, own criteria
- Pricing – raise the price to reduce demand, people can choose
- Substitutes – cheaper, often poorer quality or from different source
- Bans – difficult to enforce
- Advertising – encourages voluntary actions, types of media
- Marketing – special offers, loss leaders, etc.

2. Control supply:

- Import controls – via import tax, quotas but may face retaliation
- Subsidies – reduces costs but paid by taxpayer
- Price/cost controls – price-fixing agreements but legal problems
- Rationing/quotas – on basis of need, own criteria
- Nationalise the resource.

Exam tips

You will often be asked to evaluate the effectiveness of strategies used to manage resources. Remember that their relative success is influenced by a wide range of factors, which vary over time and space.

Factors influencing the choice and success of the methods used

- Nature of the resource – its size, type, quality, ubiquity, value, e.g. diamonds versus limestone
- Nature of the area – relief, drainage, etc (e.g. developing North Sea oil)
- Level of development/technology, e.g. LEDCs versus MEDCs
- Ownership of the resource, e.g. TNC versus state owned
- Economic factors – capital, infrastructure, power, labour available
- Social and cultural attitudes, e.g. Black Hills of Dakota, USA, were considered sacred and sacrosanct by Native Americans
- Political factors – nature of the government; totalitarian or centralised states find they can control and direct more easily
- Timeframe – short- versus long-term strategies will differ.

But, above all, demand and supply are linked.

De Beers Diamonds – South Africa and Botswana

Company created in 1880 by Cecil Rhodes during the South African diamond rush in the Kimberley area. At one time, De Beers controlled 60–80 per cent of all the world's diamonds and so was virtually a monopoly. It could control price by restricting output from its mines.

In this case, the nature of the resource aids management, as diamonds are rare, high value, low weight and very localised in origin.

De Beers now seeks to influence the output of the resource by:

- Developing new mines in joint ventures with governments, e.g. Botswana
- Developing new areas of production, e.g. in Canada
- Controlling output from its mines
- Focusing on quality rather than quantity
- Controlling the sorting, processing and cutting of the stones – 40 per cent of global output.

It tries to influence demand by:

- Using careful marketing, e.g. 'A diamond is forever' slogan created in 1947
- Placing extensive advertising in quality media
- Diversifying market, e.g. introducing the idea of eternity rings
- Guaranteeing that its diamonds are not 'conflict diamonds' (diamonds used to pay for wars and rebellions)
- Keeping prices high, so increasing exclusivity
- Having annual sales of over US$6 billion and employing over 20 000 people.

Sustainability is: enabling a resource to last in the future.

It is impossible to achieve sustainability with finite non-renewable resources so the emphasis is on:

- Making non-renewables last longer
- Ensuring semi-renewables do not go beyond the point of no recovery
- Using renewables.

Key words

Substitution
Recycling
Rationing
Benefication
Quotas

Exam tips

You need to have studied at least two different resources. Only one has to be linked to sustainability so the other would ideally be one that is finite and is becoming exhausted such as oil or a mineral like copper.

Processes that help achieve sustainability

- **Substitution** – replacing a resource with cheaper, more abundant materials, e.g. replacing copper with aluminium
- **Benefication** – improved technology, making use of previously poor-quality resource, e.g. use of old coal tips
- **Maximisation** – increasing production and reducing the waste, e.g. not flaring off gas from oil wells
- **Recyling** – reprocessing of waste to produce new materials, e.g. paper and cardboard
- **Re-use** – using old or unwanted material again, e.g. re-using cast-off clothes, building materials, etc
- **Quotas** – restrictions on how much can be used or extracted, e.g. fishing quotas in the North Sea
- **Rationing** – individuals and groups being limited in how much they can get, use or consume, e.g. food in wartime.

For case studies of attempts at sustainability, go to www.heinemann.co.uk/hotlinks, enter the express code 7719P and click on the relevant links.

Remember

Both fish and forests are semi-renewable. They mustn't be 'harvested' faster than they can be naturally replaced.

Problems associated with attempts at sustainability

Various attempts have been made to develop sustainable resources. These have chiefly focused on semi-renewable resources and have aimed to reduce demand sufficiently to allow stocks to replenish in order to ensure that demand equals supply. These attempts reflect a recognition that existing practices make long-term sustainability impossible.

However, trying to make natural resources sustainable can lead to problems because:

- Interfering in natural ecosystem processes can lead to disease, pests (as in the case of salmon farms), changes in food chains, etc
- We often don't know enough about cause and effect in natural systems, e.g. do badgers spread TB in cattle?
- Natural resources are subject to natural forces, e.g. climate change
- Human attempts at sustainability can often make things worse, e.g. the monoculture of pine trees
- There can be huge impacts on societies and cultures that rely on or use these resources
- It is difficult to achieve unless most users agree.

Two case studies follow. Fish are a particular problem because they are mobile. This makes any attempts at control difficult, as they simply move elsewhere, where the same rules may not apply.

	Common Fisheries Policy **Student book pages 150–53**	Sustainable forestry
Location	EU – North Sea	Bolivia
Aim	Make resource sustainable	Make resource sustainable 48% forested
Need	◆ Overfishing faster than fish can breed ◆ New technology 'strip-mines' the seabed ◆ Damaged or low-value fish thrown away ◆ EU treaty gave EU fishermen equal access to member states' waters	◆ Dominated by a few TNCs ◆ Unsustainable exploitation of a few species ◆ Need to expand exports from this LEDC ◆ TNCs damaging water supplies and the lives of indigenous forest dwellers
Method	◆ 1983 – policy set up ◆ 1992 and 2002 reforms ◆ Fixed quotas of allowable catches based on fish stocks – minimum mesh size. Inspectorate set up to enforce rules and fines ◆ Conservation of certain stocks and closure of some areas – recovery plans introduced ◆ Banning of certain types of fishing – promoting environmentally friendly fishing ◆ Agreements with non-EU countries over access and responsible fishing approaches ◆ Reducing capacity by scrapping old boats ◆ Encouraging sustainable aquaculture ◆ Protection of non-target species, e.g. turtles, birds, etc	◆ 1993 – BOLFOR (Bolivian Forestry) project launched (jointly with USAID) ◆ 1996 – forestry law set up monitoring scheme ◆ Concessions to develop national forests given only under strict conditions ◆ Companies that meet standards of the Forest Stewardship Council (FSC) are exempted from monitoring ◆ FSC rules introduced that cover protecting water, indigenous rights and economic well-being of workers ◆ Negotiated preferred status with ethical retailers in EU, e.g. B&Q ◆ Planned felling with re-planting ◆ Education and training on how to make forests sustainable
Impact	◆ Loss of jobs in fishing especially in smaller ports (inshore fishing protected) ◆ Fishermen see quotas as too tough, and illegal catches are dumped (= pollution and waste) ◆ Scientists see quotas as too high so some species at risk ◆ Countries disagree over their shares ◆ Loss of local fishing culture and small local ports ◆ Rising price of fish and use of a wider variety ◆ Friction with non-EU fishermen in EU waters, e.g. Russians ◆ Expansion of fish farming	◆ Expansion of fairly paid jobs in forestry (rise of 25% in incomes) ◆ World's largest share of natural tropical forests ◆ 2 million ha (25%) certified under FSC rules ◆ Exports now at £10 million a year ◆ Local economic development including ecotourism ◆ Wider range of species felled ◆ Increased habitat protection for endangered species ◆ Investment in community, e.g. health and education

ExamCafé

Student book pages 124–61

Section A

Sample question

Study resource 4 (below), which is part of an article on water resources:

Currently 96% of the world's fresh water comes from deep underground aquifers. They supply about 50% of the drinking water, 40% of water used in industry and 30% of irrigation water but this varies greatly between locations. Most are experiencing falling water levels – some 35 metres in the last 50 years. Often aquifers are shared by many countries. The Guarani aquifer is shared by Brazil, Argentina, Uruguay and Paraquay.

Identify **one** *issue and suggest appropriate management strategies to deal with it.*
[10 marks]

At first glance this question might seem unfair on those candidates who have not studied water resources or have only looked at surface water. But these questions do not aim to test your knowledge. Instead, they are designed to test your analytical skills, interpretation and the appropriateness of your solution. They are synoptic so you should draw on your wider reading and your AS work. Here there are a number of clear issues:

- We depend on underground sources for our fresh water, especially drinking water
- Water levels are falling
- Aquifers are shared resources.

Each one of these issues can be supported by evidence in the article. Do not bring in other unconnected issues which are not mentioned here, such as global warming.

Student Answer

Clearly the main issue is the falling water levels in these aquifers, which we rely on for our fresh water. At this rate there will be major water shortages in large parts of the world. One strategy to manage this would be to develop other sources of water, e.g. by desalinising seawater or using ice from the Poles. Greater efforts could be made to catch rainwater and use that to re-stock the aquifers. As water is essentially recycled via the water cycle, we need to identify areas where it is getting locked up (e.g. in products like plastics) and ensure it is released when they are recycled. More water must be re-used (e.g. bathwater to wash the car or water the garden). This is termed greywater and in new sustainable cities like Dongtan in China it is used to flush toilets and clean the streets.

This is a high level 2 response. These are logical and coherent strategies and there is evidence of synopticity (Dongtan from the AS urban unit). But it needs more depth to really look at how and why these strategies might reduce the problem (e.g. how could you use polar ice or collect more rainwater?)

Section B

Sample question

'Resources can never be made fully sustainable.' How far do you agree with this viewpoint? [30 marks]

Many of the questions at A2 follow this format – a statement against which you have to put the points that support it and those that don't.

Introductions are always important and here it is vital to set out what you understand by the terms 'resources' and 'sustainable', as different definitions could alter your discussion. The answer needs to be structured carefully so what should your paragraphs cover? Below is a sample essay plan.

Student Answer

Introduction = types of resources (renewable etc) + define sustainable

Non-renewables = oil – not sustainable

Semi-renewables, e.g. fish – can be sustainable if...

Renewables, e.g. solar power – fully sustainable

But it takes investment, technology and political will, etc

Conclusion – explore 'fully' dimension and 'can never'

This is a clear plan that progresses the argument, with both optimistic and pessimistic conclusions drawn.

One of the paragraphs contained this section:

Student Answer

To achieve even modest sustainability in resource use the public would have to be prepared to accept a cut in their standard of living or change their lifestyles drastically. We already know that in the case of simple recycling of household rubbish people will only go so far. There is a counter-argument that suggests it is best not to make resources sustainable, as the threat of imminent resource exhaustion will trigger new innovations and discoveries such as the use of cheaper aluminium to replace copper when it rose in price. Sustainability might even hold back progress, for example the shortages in the First and Second World Wars caused major advances in resources such as the invention of artificial rubber from dandelions.

This is certainly a candidate who thinks 'outside the box' and this answer illustrates the fact that there are as many ways to respond to such broad evaluative questions as there are candidates. Mark schemes are deliberately kept broad to allow the more unusual, but valid, viewpoints to gain credit. There are no right and wrong answers at A2 but to be fully effective you should finally come down on one side of the argument in your conclusion.

Chapter 5 Globalisation

Student book pages 162–205

There is a lot of useful material that supports this chapter in Chapters 4 and 6, which you might wish to look at. Some of the models of development (e.g. Rostow and core-periphery) also have implications for aspects of globalisation.

5.1 What is meant by the term 'globalisation' and why is this occurring?

Student book pages 164–71

Globalisation is: the increasing interconnection and interdependence of the world's economic, social, cultural and political systems.

Remember

Globalisation is not just about economic links – it goes much further than that.

The meaning of globalisation

Globalisation has different aspects:

- **Economic** – the interconnection of commercial and financial activities around the world. Primary, secondary, tertiary and quaternary (research and development).
- **Cultural** – the sharing, transfer and hybridisation of cultures around the world. Media, the arts, literature, food, etc.

The development of globalisation

1870–1913	Global transport systems set up, trade based on colonies supplying raw materials increased, overseas investment increased
1950–90	The rise of the TNCs setting up plants in cheap locations, exploitation of resources on a global scale
1990 onwards	Globalisation of finance, tertiary activities, spread of westernised culture and telecommunications including the Internet

Remember

There is a downside to globalisation, as the 'credit crunch' (which started in the USA) proved, when the globalisation of finances led to a knock-on effect on most financial systems around the world.

Measuring globalisation

The Centre for the Study of Globalisation and Regionalisation (CSGR) uses a number of criteria to measure the extent to which countries can be seen as globalised:

- Economic globalisation – e.g. exports and imports as a proportion of GDP
- Social globalisation – e.g. Internet users as a proportion of the population
- Political globalisation – e.g. membership of international organisations.

Top four globalised countries				
	Overall	Economic	Social	Political
1	Singapore	Luxembourg	Bermuda	France
2	Belgium	Netherlands Antilles	Singapore	USA
3	Canada	Singapore	Hong Kong	Russia
4	UK	Hong Kong	Switzerland	China

Factors responsible for globalisation

Student book page 167

Physical

- Differences in natural resources – e.g. minerals, climate, etc

Economic

- Transport revolution – cheap sea and air travel, bulk carriers
- Communications revolution – telephones, Internet, e-commerce
- Trade – cheaper to import
- Financial markets – deregulated and linked so easier to raise funds
- Comparative cost advantage – areas specialise, new international division of labour
- Economies of scale – larger markets, concentration in the hands of a few large-scale producers usually TNCs
- Ease of transferring capital – international currencies, e.g. US dollar
- Global marketing – international advertising, branding, etc

Social

- Educational and cultural – increased interest in other cultures
- Media and Internet – satellites mean instant communication anywhere
- Language – increased spread of certain languages (especially English – e.g. via the Internet) so easier communication
- Migration – ease of migration, professional mobility
- Spread of consumerism – international brands

Political

- International bodies e.g. World Bank, IMF
- Development of free market economies
- Former communist states adopting forms of capitalism
- Spread of democracy and capitalism – common links
- Removal of barriers to free trade

Exam tips

It would be difficult to single out any one factor as the most important, as so many are interconnected, but the greatest factor in establishing these global links has been the rise of telecommunications.

Key words

- Nationalism
- Isolationism
- Sinoisation (increasingly under the influence of China)
- Fundamentalism
- Protectionism

Possible future trends	
Continued globalisation	◆ Reduction in role/influence of nation states – world state ◆ Companies with no territorial identity ◆ Reduced inequalities between countries ◆ Mobile populations reacting to international demands ◆ English as common language ◆ Instant freedom of communication via the Internet
Reaction against it	◆ Rise of reactionary forces (e.g. nationalism and religious fundamentalism), isolationism (e.g. North Korea), protectionism against cheap imports (e.g. USA)
A change in direction	◆ Reduction in the importance of economic factors and development of other aspects, e.g. democracy

5.2 What are the issues associated with globalisation?

Student book pages 172–80

Remember

Even if the world decides the problems outweigh the benefits (as in the 'domino effect' of the recession), it is unlikely to be able to reverse these changes. The momentum of globalisation is too great.

Benefits and problems of globalisation

Benefits	
Environmental	Greater appreciation of 'whole Earth' approach to issues (e.g. global warming, pollution, protection of the oceans, etc), means issues can be tackled across political borders
Economic	Fall in global poverty, greater income equality, cheaper goods, greater choice of goods, less child labour, more jobs
Social	Rise in life expectancy, reduction in hunger, increased literacy/education, cultural exchange, rise of feminism, increased ease of migration, brain gain
Political	More democracy, less centralisation, more cross-border pressure groups
Problems	
Environmental	Increased pollution, global warming, destruction of environments, e.g. rainforest
Economic	Poorer countries exploited, fall in price of primary produce, low wages, local industries undercut, rise of part-time workers, financial 'meltdowns' more common as more interest in short-term gain
Social	Exploitation of workers, weaker trade unions, unskilled become unemployed, rise of underclass, culture swamped, westernisation, Sinoisation, increased illegal migration, brain drain
Political	Controlled by unelected corporate bodies (TNCs), increased centralisation on main port/capital (core versus periphery), increased dominance of consumer nations, e.g. China, USA

Case study: benefits and problems of globalisation for a NIC and a MEDC

	NIC – China Student book pages 176–77	MEDC – UK
Benefits	◆ Attracts foreign investment ◆ Rapid increase in employment ◆ Increased wages and improved working conditions ◆ Large trade surplus = investment in education, health, etc ◆ Greater travel and leisure ◆ Increased political influence ◆ Rising international profile, e.g. Olympics	◆ Attracts foreign investment ◆ Invests abroad ◆ London is now a financial hub ◆ Greater consumer choice ◆ Low-cost imports have kept inflation low ◆ More cosmopolitan society ◆ UK is now a transport and communications hub ◆ World audiences, e.g. for football
Problems	◆ Inflation – rising cost of living for poor ◆ Rapid rural-urban migration ◆ Increasing regional inequality ◆ Erosion of traditional values and culture ◆ Pressure for democratic reforms ◆ Strain on political relations with West ◆ Rising pollution and environmental damage	◆ High job losses in traditional industries ◆ Big financial and share swings ◆ Increasing wealth and pay gap ◆ Loss of political power to EU ◆ International terrorism ◆ Increased population density ◆ Rising pollution

Globalisation and the development gap

Student book pages 85–87

Remember

This is also an important part of Option 6 and there is a whole section on this in 6.3, pages 85–87.

Exam tips

You are most likely to be asked about statistical aspects in section A, where relative wealth or growth rates can be compared to see if any areas are doing better than others at closing the gap.

The development gap implies there is a gap between the wealthy MEDCs and the poor LEDCs. Currently the MEDCs have about 13 per cent of the world's population but produce 54 per cent of the output and have 85 per cent of the wealth. The table opposite shows the growth in GDP per capita for selected areas since 1950.

In the same period MEDCs increased their GDP per capita by 362 per cent but as they started from a much higher base in gross terms the growth is massive.

Is the gap closing? Those sharing in globalisation are closing the gap but those that aren't have seen incomes fall and the gap widen, e.g. in Sub-Saharan Africa. The rich get richer and the poor poorer (globally, regionally and locally). The gap between countries is narrowing but within countries it is increasing, e.g. between the rich and poor in the USA.

	1950 – GDP per capita in US$	2001 – GDP per capita in US$	% increase 1950–2001
Asia (excl China)	918	3998	435%
Africa	894	1489	166%
Latin America	2506	5811	231%
China	439	3583	816%
Eastern Europe	2111	6027	285%

Source: Adapted from UN World Economic and Social Survey 2006, *Maddison (2001) and UN/DESA, reproduced by permission of the UN.*

Quick check questions

1. Why did European countries want colonies in the 19th century?
2. When was the Internet invented?
3. Why has English become the dominant world language?
4. Which country is the most globalised?
5. Why do some cultures fear globalisation?
6. What is 'Sinoisation'?
7. Which group of companies control much of world trade?
8. What is the biggest problem of globalisation for China?

5.3 What are transnational corporations and what is their contribution to the countries in which they operate?

Student book pages 180–86

Transnational corporations (TNCs) are: very large companies with component enterprises (raw material extraction, factories, offices, outlets) in a number of countries. They market their products and/or services worldwide.

Of the top 100 wealthiest organisations in the world, 52 are TNCs.

Development of TNCs over time

Time	Type	Characteristics	Example
1500–1800	Mercantile capitalism	Government-backed chartered companies	East India Company
1800–75	Entrepreneurial capitalism	Finance houses investing in infrastructure	Railways
1875–1945	International capitalism	Manufacturers investing in raw materials	Dunlop, Cadbury
1945–60	Transnational capitalism	Investment in manufacturing in cheap locations	Volkswagen, Ford
1960 onwards	Globalising capitalism	Joint ventures and outsourcing – non-manfacturing TNCs increasing	Tesco, Wal-Mart

Contrasting organisational structures

Type of organisation	Characteristics	Example
Globally concentrated	Production at a single location but export to world markets	Canon
Host market production	Each unit serves the local national market	Coca-Cola
Product specialisation (horizontal integration)	Each unit specialises in one product for world market	Toyota
Vertical integration	Each unit carries out a stage in the production sequence	Nike
Transnational integration (Diagonal integration)	Each unit carries out a separate operation and then ships output for assembly elsewhere	Ford (Europe)

Exam tips

These different organisational structures often reflect the past history of the TNCs but also the culture of their country of origin. For example, it is interesting to compare Japanese and American TNCs.

Key words

Horizontal integration
Vertical integration
Diagonal integration

Case studies: Nike and Toyota – two TNCs

	Nike	Toyota
Product	Sportswear and equipment	Cars, forklift trucks and sewing and textile machines
Origin	1972	1918 – textile machines 1937 – cars
Head office	Beaverton, USA	Japan
Employees	65 000 – 75% in Asia and most females under 25	500 000 worldwide including dealerships
Structure	Vertical, e.g. shoes consist of 52 components from 5 different countries	Horizontal – several plants produce the same and each plant is self reliant
Location	700 factories in China, Thailand, South Korea and Vietnam	52 bases in 27 different countries
Turnover in 2005	US$15 billion	US$60 billion

Impact of TNCs	On the 'home' area/the 'exploited' area/themselves	
	Advantages	Disadvantages
Environmental	Conservation can be funded	Pollution, land clearance, loss of habitats
Economic	Jobs, higher wages, improves skill base, communications and infrastructure, multiplier effect, larger tax base, expands trade, cheaper goods	Exploits labour, low pay, destroys local industry, takes workers from primary sector, raises prices (inflation), investment can easily be moved elsewhere
Social	Education and training, health, pensions, company housing	Long hours, poor health, encourages migration to city, may undermine local culture
Political	Larger tax base, more international influence	Too much political influence (often leads to corruption), regional inequality

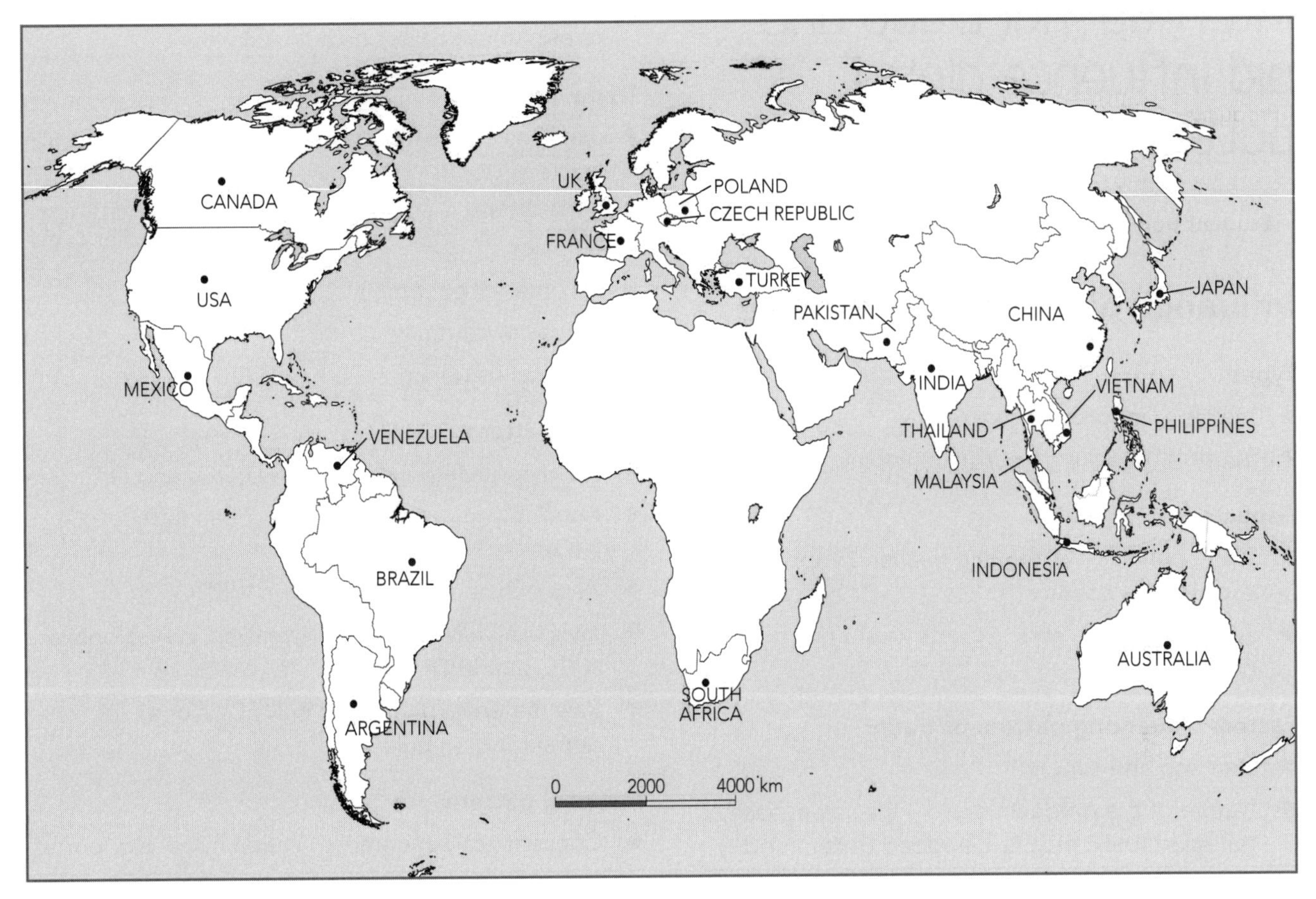

Toyota: location of manufacturing and assembly plants

> **Exam tips**
>
> If asked to evaluate the impact of TNCs, remember that the impact varies according to your viewpoint. Always consider whether there was a realistic alternative that could have achieved the same or better impact.

Example: Nike in Vietnam

Student book pages 185–86

Nike has 34 subcontracted factories in Vietnam.

Benefits to Vietnam

- Creates jobs directly, and indirectly in supplying factories
- Higher wages than local employers
- Training and education raises skills base
- Creates multiplier and cumulative causation effects
- Exports earn foreign exchange
- Sets high standards for local employers, e.g. on safety
- Feeds into local tax base
- Improves local infrastructure, e.g. power, phones, etc.

Disadvantages to Vietnam

- Exploitation of cheap labour – long hours
- Undermines local culture and practices
- Gives TNC a lot of political influence
- Competes with local employers for labour
- Profits leave the country – and go to USA
- Undercuts local producers
- Investment can easily switch to even cheaper labour sources, e.g. China.

> **Remember**
>
> In spite of their disadvantages, TNCs do provide a quick way to achieve economic take-off.

5.4 How far do international trade and aid influence global patterns of production?

Student book pages 186–98

Influence of international trade

Types

- Visible – imports, exports
- Invisible – banking, tourism, shipping, etc.

Aspects

- Trade balance – difference between imports and exports of a country
- Terms of trade – what exports would buy in terms of imports.

Factors influencing pattern of trade

- Demand and supply
- Nature of the product traded – the value, ease of transport (bulk, etc), value added of commodity
- Comparative costs of production in different areas – which in turn reflect differences in physical, economic and social conditions
- Historical factors, e.g. old colonial patterns of trade
- Political factors including trade groupings, e.g. European Union (EU), Economic Community of West African States (ECOWAS).

Trade influences

- Patterns of consumption – the variety, quality and costs of goods
- Enables areas to specialise in producing goods/services
- Incomes of countries
- Values of currency
- Political influence.

Global pattern

- Heavy polluting industries to LEDCs and NICs
- Assembly industries needing cheap labour to NICs
- Research and development in MEDCs
- Invisibles in MEDCs but increasing in Anglophile NICs, e.g. India
- Raw materials processed in MEDCs or at source, depending on bulk.

National pattern

- Concentrated at communication hubs, e.g. ports, airports
- Located in major cities, capital – skilled labour, political factors
- Accessible areas more favoured than remote = regional inequality.

Case studies: 2006 trade figures (in US$)

	LEDC – Bolivia	NIC – China	MEDC – UK
Exports	$4 billion	$970 billion	$450 billion
Agricultural	17%	3%	5%
Fuel and minerals	72%	4%	13%
Manufacturing	11%	92%	78%
Destination	Brazil 38% USA 10%	USA 21% EU 19%	EU 62% USA 13%
Imports	$3 billion	$790 billion	$620 billion
Agricultural	11%	7%	9%
Fuel and minerals	11%	20%	12%
Manufacturing	78%	73%	65%
Origin	Brazil 20% Argentina 16%	Japan 15% EU 11%	EU 50% USA 8%
Invisible trade	Exports $0.4 billion Imports $0.8 billion	Exports $91 billion Imports $100 billion	Exports $230 billion Imports $171 billion
Chief exports	Travel and tourism	Travel and tourism	Finance and banking
Chief imports	Transportation	Commercial services	Travel and tourism
Trade balance	Surplus $0.6 billion	Surplus $171 billion	Deficit $110 billion

Key words

Balance of trade | Invisible trade
Terms of trade | Trade bloc

Exam tips

You are not expected to remember the exact figures for these aspects of trade but they should be of the right order.

Organisations can influence the nature of trade patterns.

Governments seek to limit or control the nature of their trading links in order to:

- Avoid cheap imports undercutting local industries (protectionism)
- Avoid being over-dependent on foreign sources
- Increase their revenue
- Maintain employment
- Reduce wasteful expenditure on non-essential items to the minimum
- Help or disadvantage particular areas, often for political reasons
- Maintain high employment in domestic growth industries
- Safeguard trade balance and value of the currency.

They do this by:

- Imposing tariffs to make imports more expensive (e.g. the EU set import tariffs on non-EU agricultural imports of 40 per cent)
- Imposing import bans (e.g. Nigeria banned the importing of goods it could make itself such as textiles)
- Making favoured partner or preference agreements (e.g. former EU colonies in Caribbean had relief of 90 per cent of the tariff on imports of bananas)
- Setting quotas – only taking so much from a particular source
- Putting bans on exports (e.g. the USA will not trade with Cuba)
- Imposing exchange controls – limiting foreign money in the country (e.g. Mexico limits how much of its currency can be taken out)
- Subsidising exports – reducing the cost of exports (e.g. China was accused of 'dumping' cheap shoes on the European market)
- Making barter agreements – trade goods are exchanged between countries (e.g. Venezuela exchanges oil for Bolivian products)
- Setting up trade agreements between groups of states (a trade bloc) to increase trade within that group, e.g. EU, ECOWAS.

But these government strategies reduce efficiency and reduce trade.

There have been a number of attempts to create free trade on a global scale by means of trade agreements:

- GATT – General Agreement on Tariffs and Trade established in 1948. This treaty was intended to gradually reduce trade restrictions, often product by product (e.g. textiles). By 1990, 75 countries had signed up to GATT.
- WTO – World Trade Organization, set up by GATT in 1995 as a negotiating body. So far, 150 states have signed up to the WTO. There has been an increasing move towards reducing barriers that discriminate against LEDCs and including services and intellectual property rights within WTO regulations.

Key words

GATT | WTO | ECOWAS

Free trade is vital to:

- Allow countries to specialise in what they do best
- Reduce costs of goods, materials and services = higher standard of living
- Give consumers a greater choice
- Encourage efficiency and cuts in cost of production
- Allow cumulative causation and the multiplier effect to function.

The terms of trade moved steadily against countries exporting primary goods in the late 20th century, as imported manufactured goods and machinery rose in value. This only exaggerated the development gap (between the core MEDCs and peripheral LEDCs). The 21st century has seen a rise in primary produce values, compared to manufactured products, so the gap should start to close (a form of spread or trickling down).

Exam tips

Many argue that trade, not aid, is the answer to LEDCs' problems. In the long term that may be true but trade alone will not help people who are suffering the effects of drought or famine.

Influence of international aid

There are different types of international aid, which vary according to:

- **Timing** – emergency versus long term, loans versus grants
- **Source** – multilateral versus bilateral, charities, NGOs, companies, UN, etc
- **Constituents** – financial, capital goods, military, experts/technical advice, information, food, etc
- **Relationship** – former colony, military pact, part of trade bloc, etc. Is aid tied?
- **Why aid is needed** – foreign exchange gap, savings gap, technical gap, medical gap, knowledge gap.

Key words

Multilateral
Bilateral
Non-government organisations (NGOs)
Tied aid

Advantages and disadvantages of aid for recipient countries

Advantages

- 'Seed corn' for multiplier effect
- Saves lives
- Slows regional and rural to urban migration
- Leads to healthier and better skilled workforce
- Transfers knowledge without the need to discover/invent it (saves time)
- Helps bring country into world markets, trade networks, etc
- Reduces expensive imports
- Creates jobs, raises wages, etc
- Encourages reforms and improvements.

Disadvantages

- Undercuts local industry/farming
- High interest rates – drain on economy, debt burden
- Distorts local prices and incentives
- Economic colonialism
- Hidden costs (e.g. spares, fuel etc) so donor benefits
- Undue political influence
- Corruption – aid is diverted
- Increases regional inequalities
- May be wasted on 'big projects'
- Encourages growth of larger public sectors
- May delay reforms and improvements.

Exam tips

Aid is often seen as patronising by the LEDCs. MEDC aid may be inappropriate owing to cultural differences. Often, known LEDC practices are more efficient than aid offered by MEDCs, e.g. in rural India a bullock cart is more useful than a truck.

Advantages and disadvantages of aid for donor countries

Advantages

- Moral feeling of helping
- Creates sources of income, e.g. need for spare parts for machinery
- Spreads political and economic influence
- Reduces political or social instability
- Creates new markets.

Disadvantages

- Cost
- Recipient may become over-dependent
- Funds lost in corruption or misspent
- Seen by recipient as patronising
- Recipients often resent donor country.

Remember

The old expression that there is 'no such thing as a free lunch' is very applicable to aid from MEDCs.

Top aid givers and receivers in US$ (2000)

Main givers (gross)	1st: Japan $12 billion	2nd: USA $10 billion	5th: UK $4 billion
Main givers as % of GNP	1st: Denmark 1%	2nd: Norway 1%	14th: UK 0.28%
Main receivers (gross)	1st: China $3 billion	2nd: Egypt $2 billion	3rd: India $1.7 billion
Main receivers (per person)	Cape Verde $320	Seychelles $300	Dominica $270

Increasingly there is a move away from tied aid and government control towards NGOs.

Exam tips

This movement away from government aid towards NGOs is an important trend and reflects a growing realisation that such aid is more effective at the grassroots level.

Case study: Aid to Bangladesh (figures in US$)

	Short-term emergency aid	Long-term aid (NGO)
	Bangladesh cyclone, December 2007	Bangladesh Rural Advancement Committee
Amount (in US$)	Saudi Arabia $100 million, USA $2 million and UK $18 million	◆ Employs 110 000 people ◆ Income $317 million in 2007 from 11 different aid sources
Aim	To help 6 million people affected and rebuild 1 million homes that were destroyed	◆ Started as a relief agency in 1972 ◆ Now aims to give long-term basic aid to poor, e.g. health, education
Type of aid	◆ Immediate relief aid, e.g. boats, blankets, tents, clean water ◆ Medium term – to rebuild homes and businesses ◆ Also aid from charities, e.g. Red Crescent UK – Oxfam, CARE, ActionAid	◆ Operates 52 000 schools and own university ◆ Set up own bank to provide credit for small businesses – lent $4.6 billion ◆ Runs a chain of clothing stores, tea estates and dairies – $28 million sales ◆ Runs schemes to develop fisheries, poultry and forestry
Impact	Concern that aid that had to go through the government did not meet areas of need (remoter)	◆ Operates now in 5 other poor countries, e.g. Sudan ◆ Seen as more effective than government by donors – so NGOs now get 30% of all aid to Bangladesh
Direction of aid	Top down – people rely even more on aid	Bottom up – makes people more self reliant

Remember

The motives behind giving to charity to support long-term development projects may be very different to those behind giving to help an area following a natural disaster.

Example of a charity: Oxfam International

Set up in 1942 in the UK as the Oxford Committee for Famine Relief (by Quakers and academics in Oxford) to send food to starving women and children in Nazi-occupied Greece during the Second World War. Oxfam International now has 13 organisations working in over 100 countries.

Charity focuses on:

- Development – tries to lift communities out of poverty with long-term, sustainable solutions based on their needs, including long-term aid to set up basic services such as health training, clean water and schools and provide tools
- Emergencies – provides shelter (e.g. tents), clean water, sanitation, medicines, etc
- Campaigning (as a pressure group) – e.g. for better working conditions, fair trade policies, fairer rights for women
- Advocacy – human rights for those who have little power over their own lives, e.g. poor people and women.

Fund-raising (turnover of £190 million in 2003):

- Over half a million people in the UK make a regular financial contribution towards Oxfam's work – donations totalled £74 million in 2003
- Gifts left to the organisation in people's wills.
- Sales (£65 million) from its network of over 700 charity shops in UK selling secondhand goods (25% from books) and craft goods from LEDCs
- Fund-raising events, e.g. London Marathon
- Sponsorship from companies and events
- Grants from UK government, EU and UN – £40 million
- Selling its services, e.g. supporting the Glastonbury Festival as stewards.

Remember

Aid is small scale compared to the benefits of fair trade but it does transfer resources from 'the haves' to 'the have nots'. In this way, it helps to close the development gap but much depends on the type of aid and the 'strings' attached to it.

Quick check questions

1 What is the TNC structure called in which each unit of the corporation serves the national market of the country it is located in?

2 What is the obvious danger to a TNC of vertical integration?

3 Who gains from the profits of Nike's operation in Vietnam?

4 Where do most TNCs locate their research and development?

5 What is invisible trade?

6 Which import is the largest for Bolivia, China and the UK?

7 Most MEDCs have a deficit in visible trade so how do they balance their trade?

8 What is the chief item of invisible trade for most countries?

9 Why doesn't the USA (in 2009) trade with its near neighbour Cuba?

10 What is bilateral aid?

11 Why do you think small remote islands seem to get the most aid per person?

12 Why do you think Saudi Arabia gave so much aid to help Bangladesh?

5.5 How can governments evaluate and manage the impact of globalisation?

Student book pages 198–201

Measuring and evaluating the impact

Measures were referred to in 5.1 but evaluating the impact of globalisation is difficult for governments. This is because the extent and impact cannot be realistically and accurately measured, especially as so many aspects are non-quantifiable (e.g. impact on culture). Most aspects of globalisation also bring costs as well as benefits – often to differing sections of the community or economy. Some countries (e.g. the UK) see the impacts as very positive because they are hubs (transport, communications, financial, etc) in the new global system.

Exam tips

Governments tend to produce evaluations of globalisation that support their view of the phenomenon.

Different government approaches

1 Encourage (e.g. UK)

- Remove trade barriers
- Encourage international links – air, sea, Internet
- Reduce bureaucracy for foreign investors, open stock exchange
- Invest abroad
- Free market economy with no state intervention
- Tax harmonisation
- Open borders.

2 Cherry pick (e.g. Australia, Nigeria)

- Impose selective trade barriers (e.g. on products made within the country)
- Place selective controls on migrants (e.g. by wealth, type of job, etc)
- Give subsidies for selected industries/agriculture
- Stipulate minimum percentage for local ownership, indigenous managers, etc
- Nationalise key foreign investments.

3 Discourage (e.g. North Korea)

- Erect trade barriers
- Increase state control/ownership, nationalisation
- Control media including the Internet
- Control movement of people, close frontiers
- Withdraw from international treaties, trade blocs, etc
- Give subsidies for local industries – import substitution
- Become self-sufficient.

According to the old free market argument, the removal of barriers and controls brings great benefits. However, it also exploits the poorest or weakest (core-periphery model) so some protectionism may be needed.

Trying to reduce the harmful impacts of globalisation

Case study: Bolivia

Student book pages 199–201

Why manage globalisation?	◆ Fear of dominance and exploitation by USA and TNCs ◆ South America's poorest country ◆ Sense of nationalism and independence ◆ High levels of out-migration ◆ Corruption and inflation common ◆ Huge debt – over US$5 billion
Methods	
Resource nationalisation	2006 – oil and gas (Bolivia now owns 82% of production) and higher levels of royalties from companies
Resource conservation	Development of sustainable forestry – forest reserves with tight controls on felling
Trade agreements	2006 with Cuba and Venezuela (anti-USA) – exchange of soya for oil or Cuban doctors
Taxation	Increased taxation on foreign investments and companies located in Bolivia
Debt relief	2005 – debt relief scheme negotiated with G8
Land reform	2007 – 0.5 million ha redistributed to poor farmers, chiefly from large estates
Education and health	Focus on literacy and health (e.g. eyesight programmes) to raise productivity of population. Self funded rather than aid based

Exam tips

The aim should be to make the global system and its components more sustainable.

Remember

You need to consider the different models of globalisation. If all countries adopted Bolivia's approach would globalisation grind to a halt? Or is it simply that countries need to take this approach at particular stages of their development (see Rostow model)?

ExamCafé

Student book pages 162–205

Section A

Sample question

Study resource 5, which shows the components of the export trade of a LEDC in 2007.

Identify ***one*** *issue and suggest appropriate management strategies to deal with it. [10 marks]*

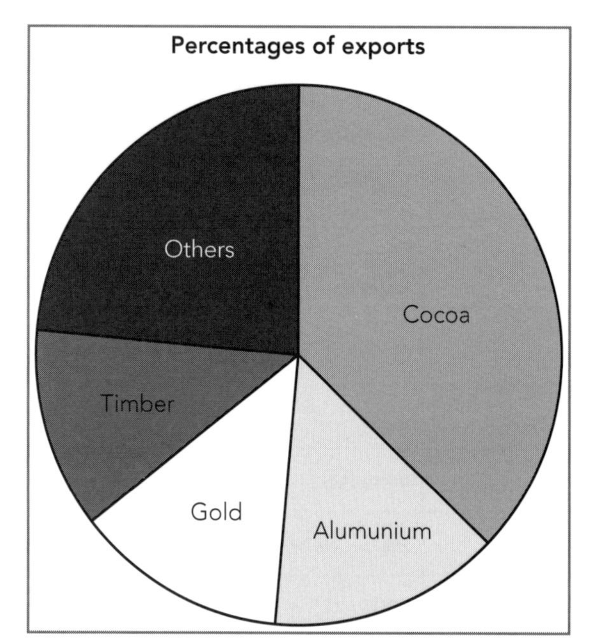

Resource 5: Percentages of exports

There are clearly a number of issues about the components of this LEDC's exports:

- There are relatively few exports so the country is at risk from changes in demand
- They are largely primary exports so the country suffers from poor terms of trade
- They are non-renewable in the case of the two minerals and semi-renewable in the case of timber
- Cocoa is vulnerable to pests, disease and climatic problems
- Apart from gold, the exports are bulky heavy items with relatively low value.

Student Answer

The country has about 35% of its export trade based on cocoa. This makes it vulnerable to competition from other LEDC producers and from natural problems such as pests or tropical storms. It should not rely on one product to such an extent. Clearly one short-term strategy would be to diversify the agricultural base to develop other tropical cash crops such as palm oil or form a cartel with other cocoa producers to force up the price to the major exporters. In the long term the strategy needs to be moving away from primary products and developing a manufacturing base, probably based on cheap labour (as it is a LEDC, it is unlikely the labour costs are high).

This is a sound level 2 answer but the student has missed opportunities to develop the depth needed for a level 3 mark. It is sound because it correctly identifies (and quotes from the resource) an issue and then provides strategies with different timeframes. This places it at the top of level 2. However, the lack of depth, in explaining how and why these suggested strategies would help the export structure, holds back the answer.

Section B

Sample question

To what extent does international aid influence the global patterns of production? [30 marks]

This is an example where the wording in the specification influences the way a question is phrased. The examiner really wanted to look at the broad advantages and disadvantages of international aid but the specification ties that issue to donors and recipients, which makes for a long or clumsy answer. So does this question demand something different?

The key thrust is on the comparative influence of aid on global patterns of production (primary, secondary, tertiary), not just the pros and cons of aid. So a conclusion might state:

Student Answer

Aid plays only a minor role in influencing the global pattern of production. There are far more important influences such as trade, FDI, technology, etc. Aid is a minor component of the forces that seek to spread development out or down from the core areas such as Europe, the US and Japan. It may help individuals and communities especially in times of emergencies but it rarely changes the global pattern of production. Grants and loans as aid have some impact but little of this is true aid and these often simply reinforce the have and have-not division of the world. Aid does seem to be 'crumbs from the rich man's table'.

This answer recognises that the question requires an evaluation of the scale, nature and impact of aid relative to other influences and the global pattern.

Another candidate made this observation:

Student Answer

Emergency aid to help the victims of a natural disaster such as the Indian Ocean tsunami would on the surface appear to have no direct influence on patterns of production at the global scale. It is given for altruistic reasons as an expression of support and sympathy. But even here it can be argued there is an underlying economic element. By helping the victims and rebuilding their lives and livelihood they return to being consumers who want the exports of the donor countries. They return to being part of the global pattern of production and consumption.

This is a perceptive and high-level comment showing that the candidate has thought through the real influence of aid. Remember that at A2 your ability to analyse, discuss and evaluate is more important than knowing the content. In this example knowing about the different types of aid would help but it is their relative role that is most important.

Chapter 6 Development and inequalities

Student book pages 206–45

6.1 In what ways do countries vary in their levels of economic development and quality of life?

Student book pages 208–16

Development is: a process of change within countries and their societies.

Remember

Development is multi-stranded and it is about much more than just economic development, but economic development usually drives progress.

Quality of life is:

- Psychological – about happiness, fulfilment, security
- Physical – diet, health
- Socio-economic – employment prospects, leisure
- Cultural
- Political – freedom, security.

Ways of measuring development level and quality of life

Quantitative – should relate to total population and be objectively measured

Demographic

- Level of birth rate (BR), death rate (DR), percentage growth
- Population density
- Diet: daily calorie supply (percentage of needs), percentage malnourished
- Health: infant mortality, life expectancy in years, percentage obese.

Economic

- Income or wealth per head, e.g. gross domestic product (GDP) per head
- Possessions, e.g. cars per 100 people, percentage with TVs/computers
- Employment: percentage in primary, percentage unemployed or underemployed
- Level of savings/investment per head
- Infrastructure, e.g. kilometres of road, percentage with phones
- Consumption levels, e.g. power consumption
- Average annual rate of economic growth.

Social

- Percentage of total income/wealth that richest 10 per cent have
- Education: adult literacy (percentage), percentage of 5-year-olds in school, percentage at university
- Percentage of teenage pregnancies, percentage on drugs
- Services: doctors/dentists per 1000 people, number of library books taken out per year
- Crime/violence: murders per 1000 people, number in prison
- Housing, e.g. percentage with indoor toilet, percentage owner occupied
- Communications: phone subscribers per 1000.

Political

- Percentage voting in elections
- Size of police force/army
- Percentage of industry that is state owned
- Number of political prisoners
- Environmental pollution, e.g. CO_2 emissions, percentage rivers polluted

Many are combined measures

- Combined Human Development Index (HDI) – combines life expectancy, adult literacy, purchasing power
- Quality of Life Index (QLI)

- Index of Sustainable Economic Development (ISED)
- The Genuine Progress Indicator (GPI)
- Gross National Happiness (GNH)
- Happy Planet Index (HPI) – this is an index that compares life satisfaction with the ecological footprint.

Qualitative – these are subjective views or perspectives that are non-quantifiable and are based on concepts such as freedom, justice, peace, happiness

Ways of classifying countries

1. First, second, third world countries
2. North–South divide
3. Development stairway
4. A continuum

Stairway groups

LDCs – least developed countries, e.g. Haiti

LEDCs – less economically developed countries, e.g. Morocco

RICs – recently industrialising countries, e.g. China

NICs – newly industrialised countries, e.g. Malaysia

MEDCs – more economically developed countries, e.g. USA

Plus:

OPEC – oil-producing and exporting countries, e.g. Saudi Arabia

FCCs – former communist countries, e.g. Russia.

Exam tips

You are not expected to remember exact figures for the measures for various countries but they should be of the right magnitude relative to other countries.

Case studies: measuring quality of life in a LEDC, NIC and MEDC in 2005

	Bangladesh – LEDC	China – NIC	Japan – MEDC
Demographic measures			
Population density (per km^2)	1045	140	339
Population growth (% per annum)	1.9	0.6	0.0
Life expectancy (years)	63	72	82
Economic measures			
GNI per person (in US$)	$470	$2010	$38 950
GDP growth	6%	10%	3%
% of population below poverty level	44	12	12
Social measures			
Daily calorie intake per person (in kcal)	2105	2844	2905
Internet users per 1000	3	85	668
% of population obese	0.03	23	25
Adult literacy (%)	34	91	99
Political measures			
Prison population per 100 000	54	117	58
Environmental measures			
Energy use (kg oil equivalent per person)	145	889	4169
CO_2 emissions (tonnes per person)	0.3	3	10
Composite measures			
HDI	0.53	0.77	0.95
QLI	5.6	6.0	7.4

Problems of measures

- Often unclear what it means, e.g. population per doctor – what type of doctor?, where located?, ignores number of visits to doctors, etc
- Accuracy of measurement is suspect in some countries, e.g. North Korea
- Hides regional variations within countries, e.g. massive difference between Delhi in India and the remote mountains of Kashmir
- Hides distribution between groups in society, e.g. gender, age, ethnicity differences
- Differences in the exchange rates often confuse things when all based on US$.

Remember

It's difficult to make exact measurements of development and quality of life. Such measures may hide important contrasts so at best they can only give a general idea of differences.

Key words

Life expectancy
Calorie intake
Gross National Income (GNI)
Gross Domestic Product (GDP)
Human Development Index (HDI)

6.2 Why do levels of economic development vary and how can they lead to inequalities?

Student book pages 216–25

Factors influencing levels of development

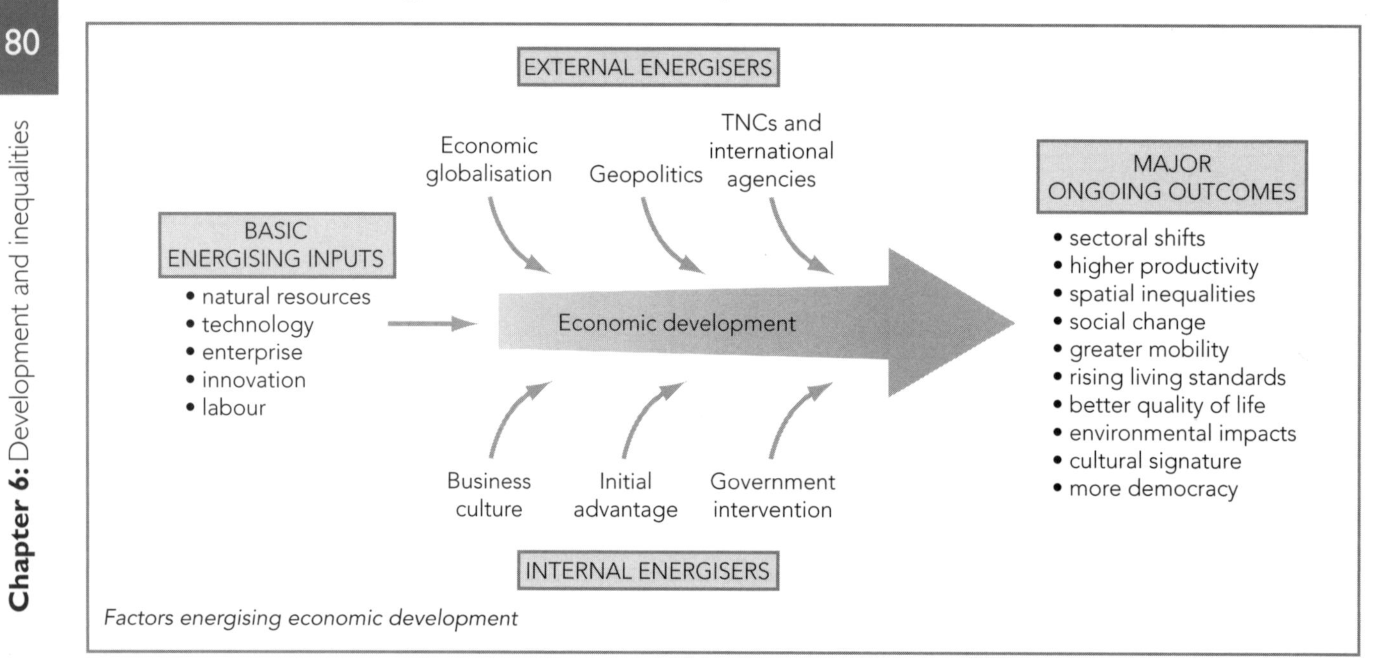

Factors energising economic development

Types of factors

- External – TNCs, aid agencies, World Bank, trade
- Internal – culture, initial advantages, government policies
- Inputs – physical, economic, social, political
- Role of history – e.g. colonialism, wars.

These impact on

- Economic, social, cultural and political systems or combinations of these.

Physical factors

- Level and type of mineral resources
- Type of climate
- Steep versus flat relief
- Prone to natural disasters?
- Water quality and quantity
- Type and fertility of soil
- Biotic resources
- Disease prone or not?

- Accessibility (centrality).

Economic factors

- Natural source of energy
- Mineral types and quality
- Building materials
- Agricultural and forestry resources
- Level of transport
- Route centre
- Port facilities
- Trade blocs.

Social factors

- Population size, type and growth rate
- Importance of and type of religion
- Level and types of education
- Standard of health
- Attitude to risk and innovation
- Level of unionisation
- Attitude to elderly.

Political factors

- Level and quality of political control and security
- Fiscal policy
- Attitude to trade and aid
- International relations
- Attitude to international migration.

Historical factors

- Colonial development/exploitation
- Role of inertia
- Former links.

Exam tips

All of these factors, often pulling in different directions, suggest that it would be impossible for inequalities not to exist. Isotropic plains (flat, featureless, uniform surfaces) do not exist in reality.

Case studies: Bangladesh and Japan – contrasting examples of development

	Bangladesh – slow to develop	Japan – economic miracle
Physical factors	◆ Lacks minerals ◆ Monsoon climate ◆ Few energy resources ◆ Low lying ◆ Floods ◆ River Ganges frequently floods ◆ Coast flooded by storms	◆ Lacks minerals ◆ Temperate climate ◆ Few energy resources ◆ Limited flat land ◆ Tectonic hazards ◆ No large rivers ◆ Long coast
Economic factors	◆ Large cheap labour supply ◆ Major rice producer – still relies on primary ◆ Exploited by TNCs ◆ Receiver of overseas direct investment ◆ Poor transport – bottlenecks ◆ Low status currency	◆ Large skilled labour supply ◆ Limited agricultural base ◆ Source of TNCs, e.g. Mitsui ◆ Makes overseas direct investment ◆ Efficient transport ◆ Undervalued currency
Social	◆ Limited education ◆ Rapid population growth ◆ Corruption ◆ Farming culture ◆ Slow to innovate	◆ Educated labour supply ◆ Low population growth ◆ Work ethic ◆ Business culture ◆ Innovative ethic
Political	◆ Muslim banking system ◆ Aid receiver ◆ Stable democracy	◆ Strict control over banks ◆ Aid donor – 'with strings' ◆ Strong democracy
Historical	◆ Former British colony and civil war with Pakistan ◆ Independent 1971	◆ US aid needed to rebuild after Second World War ◆ Never colonised

How economic development can increase or decrease inequalities

Initially development tends to increase inequalities but later reduce them.

Inequalities may be:

- Spatial – core versus periphery, disadvantaged areas
- Demographic – old versus young, male versus female
- Social – different ethnic groups, different classes, urban versus rural
- Economic – rich versus poor, type of job (primary versus secondary versus tertiary).

Also the inequalities may be economic, social, cultural and/or political or combinations of these.

Models have been used to help explain inequalities.

Models of development

Core-periphery model – Myrdal, Friedman

Various models (simplifications of reality) have been suggested to explain the processes of development and its resulting implications.

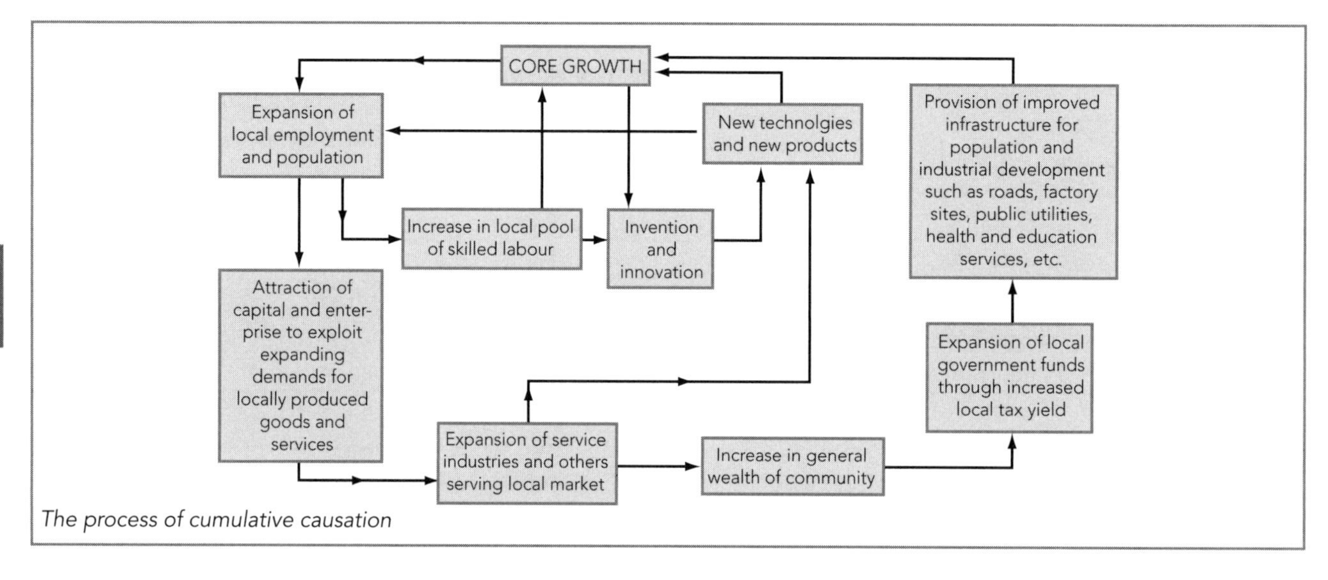

The process of cumulative causation

The core-periphery model explains why the core (or central) area grows at the expense of peripheral areas. The gap widens, as capital, labour and resources get better prices in the core and are therefore pulled there by backwash. This leaves the periphery with less educated/skilled labour, shortages of capital and poor infrastructure.

Rostow stages to growth

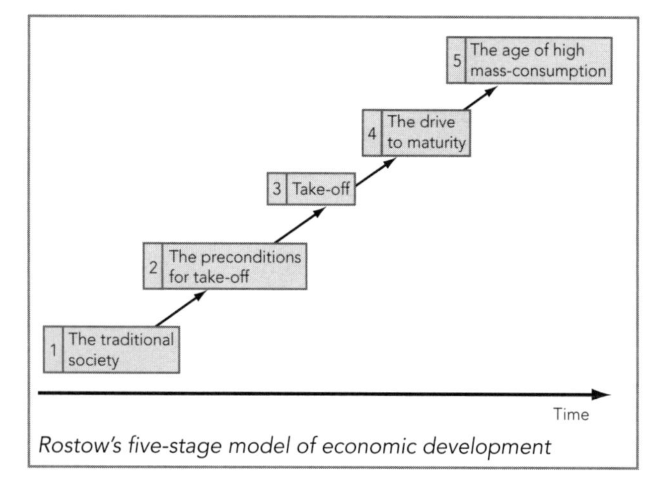

Rostow's five-stage model of economic development

The Rostow model suggests that all countries (regions) go through these five stages. The crucial one is the Stage 2 preconditions for take-off, such as exploiting resources, good infrastructure, labour supply, etc, which enable an area to achieve take-off – self-generating growth. Other areas are left behind as they lack these.

But these are free market models, based on what happened in the past in MEDCs.

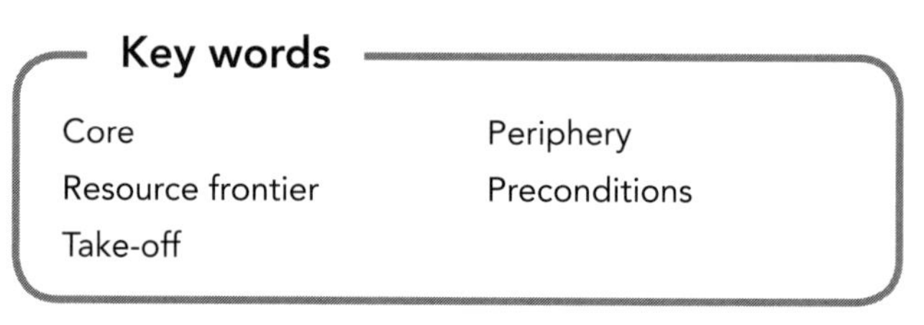

Key words

Core
Resource frontier
Take-off
Periphery
Preconditions

Key words

Backwash	Multiplier
Spread	Trickle down
Diseconomies of scale	Former communist countries (FCCs)

Factors that make inequalities increase or decrease

Inequalities **increase** because of:

- Backwash – investment, resources and labour being pulled into the core from the periphery, which thus stagnates
- Multiplier effect/virtuous spiral in core – upward spiral of wealth and employment
- Reverse multiplier/vicious spiral in periphery – downward spiral of poverty and unemployment
- Limited resources so best to concentrate them where growth potential is greatest – economies of scale.
- Urban life – better education, services, etc so birth rate falls in core but remains high in periphery
- Strong centralised political power, which concentrates limited resources and investment.

Inequalities **decrease** because of:

- Spread – core spreads its wealth, employment and demand out to remote, cheaper, less congested areas
- Trickling down – development trickles down the settlement hierarchy
- Construction of national integrated infrastructure, e.g. roads, power, Internet
- Diseconomies of scale set in the wealthy core
- Positive political decisions to reduce inequalities, possibly to prevent unrest or unpopularity – government schemes set up outside core
- New resources discovered or enabled by changes in technology, e.g. oil
- Very remoteness and underdevelopment make the area attractive for tourism.

Globalisation is reducing inequalities between countries but development tends to increase the inequalities within a country. This is because there are limited resources and these can be maximised by concentrating them where they are most productive.

Exam tips

Exam questions may refer to inequalities but remember that it is often up to you to define what this term covers.

Case study: inequalities within a country – China

Types of inequalities	◆ Poor interior versus wealthy coastal areas – coast has 50% higher per capita income ◆ Wealthy urban versus poor rural, e.g. 5% of all China's GDP is generated in Shanghai ◆ Urban Chinese earn over 3 times as much as rural Chinese ◆ Wealthy males versus poor females ◆ Industrial workers earn more than agricultural workers ◆ Rich versus poor – 204 million live on less than $1.25 a day ◆ Ethnic groups – 91% are ethnic Han but there are 55 non-Han groups (100 million population) ◆ Poor old versus wealthy young
BUT	Not just economic inequalities – there are also social inequalities (e.g. literate versus non-literate), demographic (e.g. greater female infant mortality), cultural (e.g. use of minority languages) and political (e.g. non-Communist Party members)
Causes	
Physical	◆ Coasts and river valleys favoured ◆ Interior has more arid and hostile climate ◆ Interior has more extreme relief and semi-desert ◆ Coal and oil deposits in the east and on the coast

Case study: inequalities within a country – China *(continued)*

Demographic	◆ Lower population growth in urban areas ◆ Larger families in rural areas ◆ Rapid migration to urban centres but many rural people restricted in mobility by regulations
Economic	◆ Coastal growth poles as export-driven growth – FDI (85% in east) favours already growing areas ◆ Traditional subsistence farming in interior – lack of investment in agriculture ◆ Limited infrastructure in interior – poor roads
Social	Rapidly ageing population – 1 child policy
Cultural	◆ Coastal area is more outward looking, westernised and progressive ◆ Ethnic minorities distrusted, e.g. Muslims in interior
Political	◆ Government has set up 5 special zones and 17 open cities ◆ Non-democratic and not a free market economy – so greater control of where growth is centred ◆ Rural areas still suffering from the communist land reforms ◆ Capital is in the north-eastern coastal area
History	◆ Long history of hostile interior borders or disputes, e.g. Tibet ◆ Contact with western powers via coastal enclaves such as Hong Kong

Exam tips

China is a good example of inequalities, as it has recently emerged from a communist system where inequalities were suppressed. Few inequalities should therefore exist but they do!

Attempts to reduce inequalities in China

Attempts are being made to reduce these inequalities in China because of:

- Inequalities increasing (especially rural/urban and west/east) since 1990
- Fear of political unrest
- Pollution, high costs, etc in the coastal core
- The need to increase food production to supply the growing population
- The moral and political argument of unfairness.

Economic attempts

- Development of interior resources, e.g. farming
- Improved infrastructure in the interior, e.g. Three Gorges Dam
- Spread of tourism, e.g. Xian (terracotta army)
- Decentralisation of banks and industry
- Higher food prices improve rural incomes.

Social attempts

- Poverty relief in rural areas
- Education and health schemes.

Political attempts

- Government has raised wages in state-owned rural activities
- Pensions now paid to poor and those in rural areas
- Increased military spending in border areas.

Quick check questions

1 Why do you think there are so many differing measures of the Quality of Life/Happiness?
2 Why might development not increase happiness?
3 Can you identify any group of countries missing from the stairway group?
4 Which figure in the table on page 79 suggests that China is catching up with Japan in terms of development?
5 Why is the level of population growth key in explaining the Bangladesh/Japan contrast?
6 What types of areas tend to become core areas?
7 What is the term for development moving down the settlement hierarchy?
8 Why is tourism such an effective way of creating growth in remote peripheral areas?

6.3 To what extent is the development gap increasing or decreasing?

Student book pages 225–31

What is the development gap?

- Difference in development levels between richest and poorest
- More complex than it used to be, as many countries are in between, e.g. RICs, FCCs
- Development is not just economic but includes demographic, social, cultural and political aspects as well.

Is the gap widening?

- Difficult to tell – economic and social measures differ
- Process is dynamic
- Often internal differences increase
- Some stagnate or regress, e.g. Zimbabwe.

Why are LEDCs closing the gap?

- Rapid industrialisation and cheap flexible labour, e.g. India
- Greater political stability, e.g. China
- Larger internal markets, e.g. India
- Willing to exploit globalisation – trade, communications and finance, e.g. India.

Exam tips

In an exam it is perfectly acceptable to question whether such a development gap really exists. Some would argue that it is a continuum or a ladder with many different rungs.

Case study: causes of economic regression – Zimbabwe

Growth: –4% in 2008	
Physical	◆ A string of low rainfall years ◆ Forest clearance – soil erosion
Demographic	◆ Out-migration of professionals and whites ◆ AIDS epidemic – life expectancy falling ◆ Increasing levels of malnutrition
Economic	◆ Decline in farming due to land redistribution – uneconomic small farms ◆ Rampant inflation – over 10 000 000% (September 2008) ◆ Few exports so little foreign exchange ◆ 80% unemployment
Social	◆ Slum clearance has left many homeless ◆ Inter-group tensions – insecurity and violence
Political	◆ Corruption, nationalisation of foreign assets ◆ Non-democratic – semi-dictatorship ◆ Ignored by much of Africa and MEDCs
Historical	Suffered sanctions from ex-colonial power – UK

> **Remember**
>
> Development is rarely a smooth upward curve – many areas and countries fall back or pause in their development, usually for political reasons (although natural disasters can have the same impact).

The factors increasing or decreasing the development gap

Models

- Myrdal's cumulative causation – backwash versus spread
- Rostow's stages of growth model
- Dependency theory.

Physical differences

- Relief – suitability of sites, e.g. steep slope versus flat firm land
- Natural disasters can regress a country, e.g. earthquakes
- Climate – some climates attract tourist incomes
- Drainage – water shortage, pollution, floods
- Vegetation – economic uses, loss of habitats
- Pollution – air, water, land, noise, visual intrusion
- Geology – size and type of mineral deposits, especially fossil fuels.

Economic activities

- Settlement – housing quality and quantity, cost
- Power – availability, reliability, cost
- Industry – type of jobs, lack of jobs or low pay, migrant labour, high value versus low value products
- Services – provision of schools, shops, clinics, etc, level of financial services, banking, etc
- Transport – level of connection, cost, public transport, flexibility.

Social conditions

- Wealth – relative levels, inequality and deprivation
- Cultural attitudes – view of growth, role of religion, tradition
- Age profile – dependency ratio, birth rates, growth rate, need for social services
- Migration – emigration, refugees, characteristics of migrants, depopulation versus overpopulation.

Political conditions

- Level of security – wars, terrorism
- Level of corruption
- Attitude to other countries – trade blocs, isolationism.

> **Key words**
>
> Cumulative causation
> Dependency theory
> Stages of growth
> Isolationism

Certain sectors, such as banking, are key to development.

Example: Nigerian banking

Nigeria has expanded its share of sub-Saharan banking from 14 per cent in 2004 to 34 per cent in 2008 by:

- Stamping out corruption: the Economic and Financial Crimes Commission set up in 2003 has recovered over US$4 billion and convicted over 250 people
- Making financial reforms in 1987: these reforms put in place interest rate controls, credit ceilings on loans, etc
- Increasing minimum capital required by banks: in 2004 there were 89 banks often with under US$10 million capital; now there are 24 banks that all have over US$1 billion
- Setting up micro-finance banking: in 2006, the government encouraged banks at a village level, acting as co-ops or credit unions; there are now over 700.

> **Remember**
>
> People hold different views on how the development gap is changing and this reflects their position as well as what they regard as the important elements that make up that gap.

Is the gap widening?

Viewpoints	Reasons
Reducing	Creates opportunities for all to participate, as all countries have something to offer, e.g. labour, resources
Increasing	MEDCs are in control and use terms of trade and currency values to exploit LEDCs
Fluctuating	World Bank suggests that gap widened up to 1995 and then started narrowing – reflecting more financial and political stability, rising global demand, etc
For some but not others	The rapid industrialisation of China and India has narrowed their gaps but for much of Africa the gap has grown larger

Still, 16 per cent of the world's population consumed 54 per cent of the world's output in 2005.

Global consumption shares (2000)		
Consumption	Richest 20%	Poorest 20%
Meat and fish	45%	5%
Energy	58%	4%
Telephones	74%	1.5%
Paper	84%	1.1%
Vehicles	87%	1%

6.4 In what ways do economic inequalities influence social and environmental issues?

Student book pages 232–35

Types of inequalities in society

Economic

- Fuel poverty – poor spend higher proportion on heating (over 10 per cent of income)
- Transport poverty – rely on public transport
- Food insecurity – quality and quantity
- Financial exclusion – banks, building societies, credit agencies limit funds; often have to borrow at high interest.

Demographic

- Gender – often women have fewer rights
- Age – old and/or young people may have fewer rights
- Racial – often the result of discrimination, both positive and negative
- Appearance – the current anti-obese view of many western societies.

Social

- Digital divide – fewer with Internet access
- Inverse care law – those in most need get least
- More on Social Services support
- More crime and violence – greater fire risk, vandalism
- Stigmatisation – postcode discrimination
- Poorer schools – social exclusion
- More homelessness.

Environmental

- Dereliction and decay of property, parks, etc
- Pollution – air, water, solid, e.g. litter more common.

Multiple

- Many of these interconnect to form multiple deprivation.

The link between economic inequalities and social and environmental conditions

Social and environmental issues vary between geographical areas and social groups.

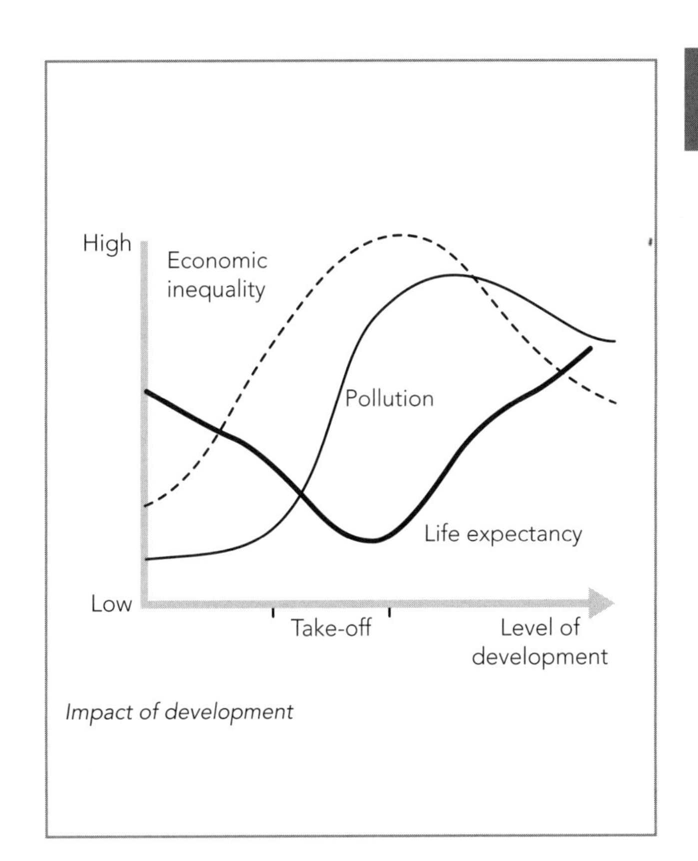

Impact of development

Inequalities in the UK

Case study: a core and periphery area in the UK

Area	Core – e.g. central London	Periphery – e.g. north-west Scotland
Social issues	◆ Rising housing costs ◆ Pressure on services ◆ Increased crime and violence ◆ Influx of young and ethnic minorities	◆ Falling house prices ◆ Loss of services ◆ Isolation and loneliness ◆ Loss of young and influx of second-home owners
Environmental issues	◆ Pollution ◆ Loss of open space ◆ Congestion	◆ Decay and neglect ◆ Conserving wilderness areas ◆ Loss of transport services
Groups	Rich	Poor
Social issues	◆ Insecurity ◆ Maintaining exclusivity ◆ Access to private healthcare	◆ Insecurity ◆ Discrimination ◆ Access to healthcare
Environmental issues	Maintaining appearance of area	Reducing hazards from the environment

World Bank environmental indicators

Remember

The evidence from Mexico and the UK suggests that environmental improvement is linked to higher rates of economic development.

Case study: a comparison of the UK and Mexico

	UK	Mexico
% of mammal species threatened	10	13
% of bird species threatened	2	6
% of land nationally protected	25	5
CO_2 emissions (metric tonnes per person)	9	4
% of population with access to safe water	100	97
% of population with access to effective sanitation	100	79

Types of environmental pollution

Types	Causes	Effects
Air	Vehicle fumes and particulates, industry, heating	◆ Increased respiratory problems, e.g. asthma ◆ Acid deposition – weathering of buildings, etc ◆ Cancer
Water	Run-off from streets, household waste disposal, sewage disposal, agricultural chemicals and waste	◆ Eutrophication ◆ Disease outbreaks, e.g. cholera ◆ Cancer ◆ Smell ◆ Damage to wildlife
Solid	Packaging, litter, 'throw-away' society, organic waste	◆ Eyesore ◆ Attracts vermin = disease ◆ Fire risk ◆ Damage to wildlife ◆ Water contamination

Noise	Traffic, domestic noise (e.g. stereos), workplace and factory noise	◆ Hearing damage ◆ Loss of sleep and increased stress ◆ Decreased concentration ◆ Nuisance – violence
Visual	Very personal view of what is ugly, e.g. wind farms, pylons	◆ Stress ◆ Psychological effects
Thermal	Buildings store heat during the day, waste heat, traffic, people, machines	◆ Creates urban heat island ◆ Increased stress
Light	Buildings, traffic, street lights	◆ Disruption of wildlife ◆ Psychological – 'loss of the stars'
Radiation	Background and also from X-ray machines and leaks from nuclear power stations	◆ Cancer ◆ Birth defects and mutations

Differences in types and levels of pollution

The type and level of pollution varies according to economic differences (e.g. traffic levels), social differences (e.g. population density) and environmental differences (e.g. relief, climatic conditions, etc).

Key words

Eutrophication
Contamination
Urban heat island
Mutation

Case studies: differences in pollution in two urban areas

	Mexico City – NIC	London – MEDC
	8th richest urban conglomeration in the world, 211th most polluted	102 of 215 in order of pollution (1 = least)
Causes		
Physical	◆ In a high altitude basin ◆ Cold air forms inversion layer	◆ Basin but low level ◆ Prevailing winds move pollution eastward (Scandinavia)
Economic	◆ Rapid industrialisation – 50 000 factories ◆ Increase in traffic – 3.5 million vehicles (many ageing)	◆ Decline of heavy industry ◆ Decline in port ◆ Traffic heavy, especially M25 ◆ London airports
Social	◆ Overcrowded – more than 6.3 people per household ◆ High population growth – more than 20 million ◆ Poor-quality housing – shanties (3 million without sewage facilities)	◆ Density is falling ◆ Housing quality is rising, e.g. slums have been cleared and people no longer use coal
Political	◆ Lax anti-pollution laws and controls	◆ Landfill taxes, EU targets ◆ Pollution laws, e.g. Clean Air Act ◆ Controls on lorry emissions with fines ◆ Congestion charge – 20% fall ◆ Recycling policy – 20%
Types		
Air	◆ Ozone levels are excessive on 98% of days ◆ Smog gets trapped under inversion layer ◆ Severe pollution	◆ Ozone levels high in summer ◆ Nitrous oxide levels exceed EU limit for 0.3 million of the population ◆ Pollution trapped in urban 'canyons' ◆ Air pollution trapped underground
Water	◆ Main aquifer polluted by industrial waste ◆ Sewage ◆ Severe pollution of water supplies	◆ Vastly reduced due to new sewage plants, decline of port and heavy industry ◆ Salmon have returned to River Thames

Case studies: differences in pollution in two urban areas *(continued)*

Solid	◆ Only 25% placed in landfill ◆ Much is left to rot = rats	◆ 70% to landfill but much exported from the city ◆ Recycling is increasing and some burnt for energy
Noise	◆ Trapped by buildings	◆ Trapped by buildings but increasing insulation and laws against excessive noise

Case studies: differences in pollution in two rural areas

	Yucatan – little pollution	Norfolk – more pollution
Causes		
Physical	◆ Very flat, largely limestone area so few rivers	◆ Very flat with porous soils
Economic	◆ Mostly forested with coastal resorts, e.g. Cancun ◆ No heavy industry	◆ Intensive agri-businesses using agri-chemicals ◆ Tourist traffic – cars and boats ◆ Increased urban development ◆ Creation of high-tech industrial estates ◆ Commuter area – traffic
Social	◆ Small population (1.8 million), largely on coast ◆ Poor area so less consumerism	◆ Rising population (0.85 million) but 9000 migrated in (2007)
Types		
Air	◆ Flat so winds blow any pollution away and few fumes are produced	◆ Worse around main roads in summer – photo-chemical smog ◆ Methane from factory farms ◆ Nitrous oxide from fertilisers
Water	◆ Water shortage means extra care taken ◆ Resorts have own sewage systems	◆ Eutrophication ◆ High nitrate concentrations ◆ Farm waste ◆ Leisure boat waste
Solid	◆ Limestone caves provide landfill sites ◆ Coast kept clean for tourists	◆ Farm waste ◆ Tourist litter
Noise	◆ Some around tourist airport	◆ Farming noise, wind farms

Economic and social inequalities in cities

Refer to the *OCR Geography AS Student book*, pages 190–94, and *OCR AS Revise Geography*, pages 54–56.

Remember

Borough statistics may obscure very marked differences between wards. Some areas of Hackney have been gentrified.

Case study: two London boroughs, 2001

	Inner city – Hackney	Outer city – Bromley
Average income	£28,600	£32,100
Unemployment	7%	2.7%
% white	59%	92%
Life expectancy (years)	64	78
Average household size	2.34	2.3

Case study: two London boroughs, 2001 *(continued)*

	Inner city – Hackney	Outer city – Bromley
No car ownership	56%	23%
Living alone	51%	31%
% retired	9%	14%
% post-16 education	32%	35%
% owner-occupied	33%	75%
% long-term illness	16%	7%

See also pages 83–84 for a regional case study on China.

Remember

Where the largest city is more than twice the size of population of the second largest city, it is termed 'primate'.

London's characteristics include:

- The largest population of any city in the UK – more than seven times bigger than Birmingham (the second city).
- Ethnic diversity – half of the UK's black, Asian and ethnic minority communities live in the capital. London's ethnic minority population stands at 42 per cent, with non-white groups making up 30 per cent of residents. The 2001 Census indicates that there were 42 communities of over 10 000 people born in countries outside Britain living in the capital.
- Age profile – London is home to nearly 1.63 million children and young people under the age of 18, accounting for almost 22 per cent of London's total population. Almost 16 per cent of London's population (1.165 million people) are aged 60 or over.
- Mobility of its population – London has many immigrants, some of whom stay for a short time. It has high numbers of refugees and asylum seekers. It also has a mobile population, with people often moving in when they are younger and looking for work and moving out when they have families.
- Large economic disparities – London is one of the world's wealthiest economies, but includes some of the country's poorest communities – 43 per cent of London's children live in households below the poverty line. Pockets of deprivation exist close to areas of extreme wealth.

Quick check questions

1. Why might Zimbabwe's figure for unemployment at 80 per cent be misleading?
2. What percentage of the world's vehicles did the richest 20 per cent own in 2000?
3. Why do you think more mammal than bird species are at risk in both Mexico and the UK?
4. Why are air pollution levels falling in London but not in Mexico City?
5. Why is the use of landfill sites not an answer to disposing of rubbish?
6. Why is Norfolk so much worse than the Yucatan in terms of rural pollution levels?
7. In the comparison between inner and outer London boroughs, what is the difference in life expectancy?
8. Is London an ageing or a youthful city?

6.5 To what extent can social and economic inequalities be reduced?

Student book pages 235–41

Remember

It is important to ask whether inequalities should merely be reduced or should be removed entirely, as a Marxist would probably advocate.

Reasons for reducing inequalities

- Moral – equal opportunities, freedom from deprivation
- Economic drain on resources, loss of potential consumers and workers
- Social – concept of a basic entitlement
- Religious or cultural – ethics of giving to the poor
- Political – reduce alienation, cut social costs, gain votes.

Methods used to reduce inequalities

- Top down versus bottom up schemes
- Government versus self-help
- Internal versus international.

Measures	Global	National
Political	Appropriate aid, cheap loans, large-scale projects	Re-development schemes, regeneration schemes, social housing, subsidies, taxation, cheap loans, improved education, improved public services and transport, better policing, anti-discrimination laws
Social	Migration, education, health (e.g. 'barefoot doctors')	Gentrification, self-help schemes, refurbishment schemes, re-branding, community development, migration, education
Economic	Direct investment, fair trade, improving international transport, appropriate aid, debt relief	Credit unions, co-operatives, improving transport

Key words

Regeneration
Gentrification
Fairtrade
Refurbishment
Co-operatives

Example: Fairtrade – Ghana pineapples

The Blue Skies Collective Association (BSOC) consists of 80 pineapple farmers who are members of four village level collectives in a poor area of central Ghana. BSOC has helped the farmers by:

- Certifying products to meet standards required by European retailers
- Achieving organic certification from the Soil Association
- Gaining Fairtrade certification
- Arranging free collection of pineapples from collection points built by BSOC
- Setting higher fixed prices for pineapples.

The farms produce 45 tonnes of pineapples a week, of which 33 per cent go to BSOC. Members rely on pineapple sales for 85 per cent of their incomes. Exports doubled following Fairtrade certification. Retailers' premium for Fairtrade/organic products goes into a central fund and is used for community projects – e.g. digging boreholes to offset water supply problems, building public toilets and providing local schools.

For more information on this project, visit the Fairtrade website. Go to www.heinemann.co.uk/hotlinks, enter the express code 7719P and click on the relevant link.

Remember

Inequalities exist in both LEDCs and MEDCs, often for the same reasons. Most MEDCs have taken action to reduce such inequalities so they are not as extreme as in LEDCs.

Case study: policies to reduce spatial inequalities in the UK

Types of spatial inequality	1. Regional, e.g. declining versus growth (crudely North versus South) 2. Urban versus remote rural 3. Inner city versus outer urban fringe
Causes	◆ Growth in south-east around London and near EU core ◆ Decline of old heavy industry and mining in coalfield areas ◆ Mechanisation of farming and depopulation of rural areas ◆ Urban sprawl ◆ Rising incomes leading to counter-urbanisation

Case study: policies to reduce spatial inequalities in the UK *(continued)*

Policies	
'Stick' to limit growth in key areas	◆ Curb growth of urban areas using green belts, e.g. London ◆ Set higher rates of tax in prosperous or growth areas
'Carrot' to encourage spread forces	◆ Establish regional growth poles ◆ Set up activities with large multipliers
Grants	◆ EU regional development policies (Euros 9b 2007–13): Objective 1 funds to poorest areas (e.g. Cornwall, north-west Scotland); Objective 2 to support regions facing structural and competitive difficulties (industrial and rural), e.g. Tyneside ◆ Rural Development Areas and Rural Challenge Scheme ◆ Assisted Areas and Enterprise Zones ◆ City Challenge
Tax incentives	◆ Tax breaks for those locating in declining areas ◆ Rate relief for poorer Local Authorities
Infrastructure	◆ New roads, regional airports, ports ◆ Targeted strategies, e.g. Rural Transport Partnership Scheme
Construction	◆ New and expanded town policy ◆ Establishing eco-towns
Relocation	◆ Offices, e.g. tax offices to north-east, Driver and Vehicle Licensing Authority (DVLA) to Cardiff ◆ Military bases
Development projects	◆ New power stations, e.g. Severn Barrage ◆ Tourist developments, e.g. Eden Project in Cornwall
Urban redevelopment	◆ Urban regeneration schemes, e.g. Manchester ◆ New Deal for Communities ◆ Garden festivals
Political	◆ Regional devolution, e.g. Scotland

But market forces have done more to reduce inequalities, as people have moved out of high cost areas to cheaper locations, enabled to do so by increased mobility. Should governments interfere in the free market mechanism?

Methods governments use to reduce personal inequalities

- Setting minimum wages
- Passing laws on nature and conditions of employment
- Taxing wealthier people more
- Laws on equal opportunities and anti-discrimination, e.g. race, age
- Social services – health and education available to all
- Social security and support for needy, e.g. pensions
- Subsidised housing for those on low incomes.

Key words

Growth poles
Regeneration
Eco-towns
Infrastructure
Devolution

Quick check questions

1 What is a credit union?
2 Self-help schemes have been widely used to help what type of group in the community?
3 Fairtrade has what impact on the price of products from LEDCs?
4 What is the difference between 'stick' and 'carrot' policies?
5 Which model is the basis for the concept of growth poles?
6 Why might the expansion of regional airports help reduce regional inequalities in the UK?

ExamCafé

Student book pages 206–45

Section A

Sample question

Study resource 6, which shows census data for selected wards in an urban area in the UK (2004).

Resource 6:

Ward	% of white British	Higher managerial and professional	Long-term unemployed
Bradshaw	97	828	213
Burnden	70	402	660
Derby	45	253	1220
Halliwell	73	327	643
Smithills	92	640	220

Identify ***one*** *issue and suggest appropriate management strategies to deal with it. [10 marks]*

As in all these section A questions, there are a range of possible issues. In this case they could include:

- Derby ward seems to have a very low proportion (in comparison) of white British people
- Likewise, Derby ward has a very low number of higher managerial/professionals (higher income groups)
- And high long-term unemployment
- Combinations of these.

Student Answer

The table shows that Derby ward has a much higher number of unemployed (at 1220, almost double the next highest). This may be a direct result of the high proportion of ethnic minorities (55 per cent), who find it more difficult to get jobs.

This answer is sound and scores well, as it refers to the data in the resource. It may be tempting for the candidate to go on and explain why this is so but that would waste time as it would gain no credit.

Student Answer

The problem can be managed by using a range of strategies, as it may reflect the fact that the ethnic minorities are relatively newly arrived so need education or training to reduce the language barrier and upgrade their skills to fit the local labour market. Cheap grants and loans could be given to encourage the creation of small businesses supplying the needs of the local community, especially if this is supported by the building of small nursery factory units that can be rented at preferential low rents or with exemption from local taxes.

This is a good response as far as it goes. The candidate has used the cause-effect approach to justify the strategy of education, etc. In this way the student demonstrates knowledge and understanding of urban inequalities. Then a more conventional approach to reducing unemployment is developed. The two parts together produce a low level 2 response. To reach the top of level 2, greater depth would be needed in explaining why these measures reduce unemployment at the ward level. An alternative approach would be to look at other strategies such as regeneration schemes, credit unions/co-operatives, etc but there is a risk that these would add no more depth (and so gain no more marks).

Section B

Sample question

How far do you agree with the statement that 'Development inevitably increases inequalities within a country'?

Remember to read the question carefully. Key elements here include:

- 'Development' so not just economic but could include social, political, etc
- 'Inevitably' – clearly it is not 100 per cent guaranteed
- Remember the idea that inequalities first increase then decrease
- 'Inequalities' – these could be regional, rural versus urban or between different groups in the community. Are they economic, social or political?
- 'Within' so it should be securely anchored in a country **but** to develop the evaluation you may need to refer to more than one country to show a contrast, i.e. one where it has increased and one where it has decreased.

This clearly generates a structure for the essay as shown in the plan below:

Student Answer

Introduction – definition of development and inequalities

Theory of why inequalities increase – Myrdal

Example of Brazil – coastal growth

But – idea of spread/trickling down

Examples of where this is starting in Brazil

Some inequalities are inc and some dec (Ec, Soc, Polit)

Conclusion – depends on viewpoint

Chapter 7 Geographical skills

Student book pages 248–99

Remember that the specification requires you to have carried out a geographical investigation. This may be undertaken at AS and A2 and can be:

- Research using largely secondary sources
- Fieldwork (often referred to below as 'in the field').

But in either case the six stages must be followed. Each stage is summarised below, and areas or topics that are likely to form the focus of questions are picked out in more detail.

Exam tips

Fieldwork has a lot of advantages over classroom-based investigations, e.g. risk assessment is limited in a classroom.

Remember

Section B is very much about what you did as individual research so make sure it is personalised even if it has been undertaken as a group project.

7.1 Identify a suitable geographical question or hypothesis for investigation

Student book pages 250–52

Choose a topic:

- What do you want to investigate – where, when, how and why?
- It should have a clear aim.

The investigation should be:

- Suitable in scale, size, area
- Capable of research – practical issues, e.g. time, area, data available
- Clearly geographical – strong sense of location, where, pattern, etc
- Based on geographical theory, ideas, concepts, models (clear spatial or locational focus)
- Logical – it must make sense, especially cause-effect
- Clearly located.

The title should be:

- Simple, Measurable, Achievable, Realistic, Timed (SMART)
- Include a question or hypotheses (not too many).

The investigation should be based on:

- A geographical theory, e.g. distance decay
- A model, e.g. Burgess
- A concept, e.g. succession.

Choosing a title

Let's look at two sample titles.

1. To investigate the land use zones of Bristol

This is a poor title, as it is:

- Too large an area to be practical in time or area
- Not a question
- Too vague – no suggestion of a pattern or trend to investigate or test
- Doesn't give a clear outcome – no real logical conclusion.
- Apart from land use data, what else would be needed?

So is it SMART?

S = yes, M = unclear, A = doubtful, R = no, T = no.

2. To test the hypothesis that there are concentric zones of land use around the CBD of Witney

This is a better title, as it is:

- A smaller area so it's practical but still large enough to have a range of land uses

- A clear question
- Includes a clear reference to a pattern (underlying concept of Burgess model)
- A hypothesis – testing gives a clear conclusion.

Here it is land use with distance from CBD so there is a clear data collection strategy.

So is this SMART?

S = yes, M = yes, A = yes, R = yes, T = no.

Stage 1 is the most important stage, as a poor title makes all the other stages so much more difficult. Think of the data you would need to collect to test or investigate your title. Always keep titles simple and direct. Never speculate (e.g. What would be the effect of building a new sports centre in Chipping Norton?) as this will make it impossible to find your data.

Exam tips

One of the common ways of examining your ability to suggest appropriate investigations is to refer you to an OS map extract or photograph and ask you to suggest a possible investigation that could be carried out at the location or area shown.

Remember

It is often easiest to identify physical geography investigations. You can always compare and contrast two locations (such as highland/lowland, coast/inland, flat/steep, north facing/south facing, etc) in their microclimate, soils or vegetation.

Hypothesis testing

Testing a hypothesis is a good approach to designing a question for investigation

Exam tips

If you use more than one hypothesis (and there is no need to), make sure they follow on from one another.

You can **justify** your choice of title by:

- Locating it in the chosen area (say why it could be carried out in or at that location)
- Estimating the time it will take
- Estimating its scale – to make sure you can do it with the resources (equipment, team, etc) you have
- Showing its relevance for the chosen area (e.g. why it might vary there)
- Linking it with your desire to test a particular model or concept in the field.

Your **aim** is then to test or prove that title. You may want to break down the stages you would take to do this into individual objectives because these can be checked off as you complete them. Here's an example.

Aim: to test that pebble sizes on the beach decrease downdrift

Objectives:

- To measure 30 pebbles selected randomly at each of 10 locations along the beach
- To record the size of their long axis
- To plot the length of their long axis against distance along the beach.

Key words

Investigation	Hypothesis
Justify	Aim
Objectives	

Remember

Even if you don't choose your own question for investigation or title (because it has been chosen by your teacher or field studies adviser), you must understand what it means, especially in terms of its implications for your own research.

7.2 Develop a plan and strategy for conducting the investigation

Student book pages 253–58

Developing a plan

Work out what you plan to do:

- Draw up a timed or sequenced plan
- You may modify this in practice.

Work out the location:

- Where are you going to carry out your investigation?
- Show it clearly on an annotated map.

Identify the data needed:

- What are your sources of primary and secondary data?
- How much data do you need?
- What access do you have to data sources (e.g. the Internet)?
- Primary – data collected first hand; unprocessed
- Secondary – data from published sources (e.g. the Internet, libraries, previous investigations); processed
- Always specify sources and check reliability, dates, etc.

Work out your strategy for collecting data:

- Intended method (e.g. questionnaire), type of sampling strategy, location, timing, division of labour, equipment needed, timeline.

Choose your sampling type:

- Units = linear, area (quadrat) or point
- Random – statistically most useful
- Stratified – use when data occurs in clear subsections
- Systematic – lazy way, as quick and easy
- Pragmatic – sample where you can
- Can combine, e.g. find ends of transects (linear sampling) randomly, then sample systematically.

Recognise limitations:

- In terms of time, location, coverage, equipment (type, reliability and accuracy), resources, labour force available, access.

Identify risks:

- Potential risks, their seriousness and likelihood and possible ways of reducing or managing these
- Also risk that data may be unreliable or inaccurate.

Arrange a pilot:

- Need to carry out a pilot or reconnaissance to check viability of data collection strategy in the time and location chosen, e.g. a trial questionnaire.

Exam tips

Risk assessment is a common question. Fieldwork investigations, especially those that involve some physical geography, probably involve more obvious risks than secondary research.

Assessing the risks

These are risks to the person (health and safety) but also to the reliability and accuracy of the data. They stem from:

- The environment or area the investigation is being undertaken in, e.g. slippery, traffic, water depth, etc
- The time of day/year when it is being carried out
- Possible extreme weather, e.g. sunstroke
- The nature of the activity you are planning to undertake
- The methodology, including the equipment, being used
- Personal issues, e.g. asthma sufferer, inappropriate clothing
- Personnel who are working together
- Legal restrictions, e.g. rights of access.

Before undertaking any practical fieldwork activity, you should always carry out a risk assessment.

Any steps you take to reduce these risks should be appropriate and avoid unrealistic solutions. It is useful to include phone numbers of key people in any such advice.

Remember

There is often a tension between keeping risks to the minimum and making individual and accurate measurements. Today it is deemed too risky for an individual to do their research on their own (thus standardising the data collection). Data is usually therefore collected in or by a group.

Making a plan

Plans are useful guides to the expected time needed and may identify bottlenecks in the sharing of equipment, access, etc.

There are two types of plan:

1. An overall plan that looks at the stages of the investigation, probably identifying stages 4 and 5 as time consuming. This is useful in setting out the logistics, sequence of activities and how the various stages fit into the time allocated.
2. There is also the plan you draw up to organise the practical work in the field.

This is a sample field plan:

Task		Who?	When?
1	Organise team and allocate tasks	Me	8.30
2	Check equipment	DC	9.00
3	Lay out transect line	PB	9.00
4	Team members do systematic sample	All	9.15+
5	Share results	All	10.00
6	Collect and check equipment	DC	10.30

This type of plan is useful, especially if an extra column is added for comments or actual timings. This makes it into a learning tool for the future. In the example above, probably insufficient time has been allocated for the actual sampling. This often takes longer than you expect.

Primary or secondary data?

This is a crucial decision – here are some of the advantages and disadvantages:

	Primary	Secondary
Positives	◆ Real – on that day ◆ Located in your area ◆ You know how it was collected ◆ You 'own' it	◆ Accurate (as often larger sample or repeated) ◆ May give the average or norm
Negatives	◆ May not be accurate (in collection or recording) ◆ Once-off data	◆ Finding sources ◆ May be biased ◆ May be out of date ◆ May be average data ◆ Don't know how it was collected ◆ May not be exactly the same units or areas as yours

Exam tips

You can expect to be asked about sampling in most exams as it occurs in both stages 2 and 3. You must be clear why you chose, or would choose, that particular sampling strategy (size, unit and type) rather than another.

Designing questionnaires

Questionnaires are a common tool but are often poorly designed. Some points to look out for:

- Are questionnaires numbered and dated (for future reference)?
- Is the location where they are being used clear?
- Questions about names, addresses, incomes, sex, etc are too intrusive so people won't answer them
- Open-ended questions make categorising data difficult later
- If you use a series of categories as possible answers, always have one that is 'other' to cope with answers outside your groups
- Be wary of asking what people 'usually' or 'regularly' do, as most people forget, so ask what they last did
- Avoid ambiguous or unclear questions that you have to explain
- Don't ask too many questions for the time (people won't stop long)
- Is there a thank you prompt at the end?

Questionnaires are usually asked face to face but they can be posted in one form or another, e.g. with a reply envelope. Here are some of the advantages and disadvantages of each method:

	Face to face	Posted
Positives	◆ You can explain any questions ◆ You can see the sex/age of the respondent ◆ You can alter aspects as you go if they aren't working	◆ More objective, as respondents can take their time ◆ Avoid personal issues and bias ◆ Can be longer ◆ Less time constrained
Negatives	◆ Respondents like to please so they give the answers they think you want ◆ Often biased, as only certain people stop for you (e.g. males and females differ) ◆ Boredom after asking the same questions repeatedly	◆ Cost ◆ Very low response rates (less than 20% is the norm) ◆ Must be very clear, as you won't be there to explain the questions ◆ People don't like the intrusion so often throw them away

Key words

Data
Secondary data
Risk assessment
Primary data
Sampling

Quick check questions

1. All titles or questions for investigation should be SMART. What does this mean?
2. Which is the most difficult element to achieve of your answer to question 1?
3. Why should you never carry out fieldwork by yourself?
4. Suggest a risk factor when carrying out an investigation based on secondary sources.
5. What tends to take longer in reality than you expect when planning an investigation?
6. Is census data primary or secondary?
7. Why not use the Internet to send questionnaires to people?

7.3 Collect and record data appropriate to the geographical question or hypothesis

Student book pages 258–64

Collecting data

Primary data

- How did you collect the data? Equipment, strategy and methods used – their accuracy and reliability
- Questionnaire design – type of questions used (closed versus open), response approach (e.g. 4 versus 5 scale), length, postal versus face to face
- Sampling size and type – systematic, random, stratified, pragmatic
- Sampling unit – area, point, linear
- Map of sample sites, repeated sampling for accuracy.

Secondary data

- Source
- Date
- Reliability and accuracy.

Recording

- Nature of tally sheet
- Any problems.

Geographic Information Systems (GIS)

- Covers a wide range of ICT programs
- Primary data collection, e.g. data logging temperatures at specific locations/times, measuring distances/areas
- Source of secondary data, e.g. satellite photos, layers of data from maps.

Sources of data

Main ways of collecting primary data:

- Measuring and counting
- Mapping and plotting
- Making images, including sketches and photos
- Asking questions
- Using ICT – data logging.

Main sources of secondary data:

- Graphical sources – maps, diagrams, etc
- Images – photos, satellite images, Internet
- Written sources – newspapers, books
- Oral sources – radio, tapes
- Statistics – census data.

Difficulties of sampling

Sampling is often more difficult in reality than when planning it. This reduces the **reliability** of samples. Questionnaires are particularly difficult to ask randomly, systematically or in a stratified way because:

- They take time so only a limited number of people will be able or willing to answer them
- Many people are put off by questionnaires and/or students
- Certain groups find it difficult to respond to questionnaires, e.g. parents with young children
- Certain groups respond too readily to questionnaires and may positively seek them out, e.g. pensioners.

So, in reality, questionnaires are usually pragmatic – and rarely answered by a genuine cross-section of society.

Also, the nature of the population being sampled will change according to:

- Time of day
- Day of the week
- Time of year (e.g. winter versus summer)
- Other factors (e.g. the weather).

This produces biased sampling but in some cases that is what is needed, e.g. when researching traffic conditions at rush hour.

Maps and photographs

These differ in their uses. Here are some of their advantages and disadvantages:

	OS map	Photograph
Positives	◆ Accurate located detail ◆ Scale allows precise measurement ◆ Relative directions clear ◆ Can see over hills, etc ◆ Get plan view ◆ Gives placenames, etc	◆ Captures an instant of time – shows ephemeral features, e.g. weather ◆ Shows precise uses, e.g. types of tree, shop, etc ◆ Shows people ◆ Shows qualitative features
Negatives	◆ Static picture – shows only static features so can't show seasonal changes, weather, etc ◆ Quickly dated ◆ Impersonal ◆ Sheer size	◆ Instant of time – so quickly becomes out of date ◆ Scale problems ◆ Subjective – image changes with light, angle, etc ◆ Can't show 'hidden' areas ◆ Doesn't show precise classification, e.g. roads

Remember

This stage is ideal for demonstrating what you did individually to help the investigation.

Exam tips

If a question refers to modern technology it is trying to get you to refer to GIS but if your centre can't access this then other technology is acceptable, e.g. data loggers.

Field sketches

Field sketches can be useful in terms of:

- Showing a context for the data collected
- Identifying particular features
- Conveying a subjective impression of a place.

Such sketches should always have:

- A clear location – grid reference
- An informative title
- A note of the date/time
- An indication of scale (vertical and horizontal) and direction
- Annotations to identify specific features.

It is often very effective to put a photograph and an annotated sketch of the same view on the same page, or use the sketch as an overlay.

Key words

Accuracy
Geographic Information Systems (GIS)
Reliability

GIS

The advantages of using GIS include:

- Handles large amounts and varieties of data so it can be manipulated more easily
- Flexible over size of study area
- Gives precise data for precise locations
- Gives precise measurements of area, distance, direction, etc
- Easy to keep up to date as it is dynamic and therefore current
- Faster and more efficient at handling data
- Can easily change the scale or sub-divisions of data
- Can overlay patterns, so suggesting possible links
- Produces striking visual impacts, including three-dimensional images.

7.4 Present the data collected in appropriate forms

Student book pages 265–76

Presenting data

Types of data

- Discrete, ordinal, continuous, areal, time-series, period
- Qualitative versus quantitative.

Aspects of presentation to be considered

- Type of presentation (e.g. maps, diagrams, graphs, photos, charts) should fit type of data (e.g. continuous data is best shown with a line graph)

- Use of colour or shading
- Size of diagram or symbol – scale is crucial
- Ease of comparability (e.g. same scale)
- Annotation – location, detail, etc
- Lettering – font, spacing, size
- Located on a map or not
- Add a key, scale, title and north if using a map
- Logical organisation to help analysis (e.g. number any diagrams).

Types of displays

Non-spatial displays

- Tables
- Charts – bar, pie, rose, proportional symbols
- Graphs – line, scatter, triangular, compound, positive and negative.

Spatial displays

- Maps – think scale, projection
- Isopleths – think interval, interpolation, curve
- Choropleths – think boundaries, shading, class interval
- Others – trend surfaces, topological
- Located symbols, e.g. dots, proportional circles – think of scale/size, location, key.

Movement

- Flow lines, trip lines – width, direction, arrows, sub-divisions.

Photographs

- Types (landscape, oblique, aerial, satellite) – think scale, direction; remember they capture an instant of reality
- Maps show static long-term features.
- Always include scale (e.g. person, building, etc), time of day/year taken, title
- Annotate – arrows to selected items of interest and minimal text.

Exam tips

This is a popular stage to use in section A questions, as so many candidates make careless errors with their choice of presentational methods.

Selecting a presentation method

When selecting the method of presentation, consider:

- What the data type is, e.g. percentages
- The purpose of the diagram, etc – who is its audience?
- How accurate must the plotting of values be?
- The range of values in the data set
- The space available for the diagram, etc
- The visual impact desired
- Does it need labelling or annotating?
- The time you have available
- Do you need to calculate anything?
- Should you use ICT to produce the diagram?

Each method of presentation tends to have advantages and disadvantages, as shown below.

Methods	Advantages	Disadvantages
Bar chart	◆ Quick and easy ◆ Very visual	◆ Should only be used for interval data
Pie chart	◆ Very visual ◆ Easy to compare	◆ Must use percentage data ◆ Too many divisions confuse the eye ◆ Need to standardise order of segments ◆ Everyone uses them
Line graph	◆ Good at showing trends or patterns (and anomalies)	◆ Should only be used for continuous variables, which are rare
Dots	◆ Simple ◆ Shows clear location of value	◆ Difficult to see/count ◆ Is exact location known?
Isopleth	◆ Helps to show or identify patterns from individually located values	◆ Interval size is vital ◆ Interpolation is dubious or guesswork ◆ Shape is guesswork
Choropleth	◆ Quick and easy ◆ Very visual	◆ Interval value is critical ◆ Treats area as a whole ◆ Shading issues
Flow lines	◆ Shows movement along routes – direction, volume and distance ◆ Can be subdivided to show characteristics	◆ Following exact route can be over-complex ◆ Can mislead if it doesn't follow exact route

Exam tips

Bar charts are rather simple but sometimes they are the best way of showing the data. Don't look for a complex technique when a simpler one will do the job.

There are some common problems to watch out for with all the techniques:

- Scale (check that it starts at 0)
- Size of diagram
- Data interval
- Data class division
- Use of colours or shades
- Labelling.

Remember

You often have a choice of techniques to represent your data. Be consistent. Choose one appropriate type and stick to it, using the same scale/key, so that comparisons of your data can easily be made.

When you look at a diagram, check to see if it has:

- A title and figure number/letter
- A sensible scale or key
- A note of its location
- Shading or colour – are they appropriate?

Quick check questions

1 What is the chief advantage of a photograph over a map of the same area?

2 Apart from annotations and a title, what should all field sketches have?

3 When using a large-scale map, why is projection an important consideration?

4 What o'clock should all pie charts start at?

5 In a pie chart which sectors should you draw in first?

6 Why shouldn't you use colours on diagrams you draw in the examination?

7 When would the use of two distinctive colours be a good idea?

8 What is the chief problem when drawing isopleths?

Key words

Qualitative data	Quantitative data
Located symbol	Choropleth
Isopleth or isoline	Chart

Examples from the student book

A useful way of looking at the effectiveness of ways of representing various types of data is to look at examples in the student book.

1 Page 30, Figure 1.28. This map uses located dots and circles. What is its chief drawback?

2 Page 16, Figure1.14. Why is a line graph the most effective way of showing this data? Why would an average line be misleading?

3 Page 95, Figure 3.11. Why use 'decade' bars rather than year bars?

4 Page 128, Figure 4.6. Why does the line for Bangladesh start around 1940?

5 Page 136, Figure 4.19. Why have figures been added to the bars?

7.5 Analyse and interpret the data

Student book pages 276–92

Analysing the data

Qualitative

- No use of statistics
- There may be no need as pattern is so obvious – e.g. scatter graph, compare graphs.

Quantitative

- Use quantitative to be objective, know exact level or direction of relationship/trend, know exact pattern, identify anomalies
- Use the analysis to go further.

Types of quantitative data analysis

Descriptive

- Descriptive techniques – central tendency, e.g. mean, mode (most common), median (middle value), scatter graphs

- Deviation from the central tendency – range, interquartile range, standard deviation
- Frequencies (graph) – kurtosis, skew.

Normal distribution

- Mean, mode and median coincide – 95 per cent of values within two standard deviations of mean.

Types of data pattern

- Horizontal, e.g. nearest neighbour – clustered, regular or scattered
- Vertical, e.g. Lorenz curve – cumulative frequency
- Networks – beta index, alpha index, centrality
- Interactions, e.g. gravity model.

Tests

- Statistical test for difference, e.g. Mann-Whitney, Chi-squared
- Statistical test for correlation (linkage) or association, e.g. Spearman's rank, Chi-squared
- In tests remember null hypothesis and degrees of freedom influence the result and 95 per cent is the minimum expected accuracy. This is found from tables of significance.

Interpretation

- Look at results in context of original question(s)
- Try to offer explanation for patterns/links/trends and any anomalies.

Exam tips

You will not be asked to carry out a statistical test in an examination but you will be expected to be able to interpret the meaning of its result.

Remember

Most research insists on a 95 per cent accuracy level so there is 5 per cent error. If your result comes out at 75 per cent then it still shows a trend but it is not secure.

Types of statistics and their uses

Measure	Its meaning	Its limitations
Central tendency and dispersion		
Mean	Add the quantities together and divide by the number of quantities	Distorted by extreme values and involves calculation
Median	The central value when values are put in order	Gives no idea of other values
Mode	The value that occurs most frequently	Gives no idea of other values
Range	Difference between highest and lowest values	Distorted by extreme values
Interquartile range	Difference between the inner half of the data around the median	Ignores values above and below this
Standard deviation	Shows spread of all values around the mean	Uses a formula and calculation and distorted by extreme values
Tests for differences		
Mann-Whitney	Compares medians and ranks to see if data set differs	Uses a formula and calculation, and uses a significance table
Chi-squared	Compares observed and expected frequencies	◆ Uses a formula, calculation and significance table ◆ How do you determine expected frequency?
Tests for association		
Spearman's rank	Measures strength of relationship between two sets of ranked data	◆ Uses a formula and calculation ◆ Only uses ranks of data and uses a significance table
Chi-squared	Compares observed and the frequency expected given a certain hypothesis	◆ Uses a formula and calculation and significance table ◆ How do you determine expected frequency?

Key words

Clustered
Regular
Degrees of freedom
Significance
Random
Correlation
Null hypothesis
Normal distribution

Trends and relationships

These are often shown on graphs, especially scatter graphs and tell you:

- The direction of the trend/relationship – is it positive, negative or neutral?
- The strength of the trend/relationship – is it strong, weak or non-existent?
- The shape of the trend/relationship – is it linear, parabolic, exponential, unclear?
- If there are values that are anomalies (don't fit the trend/relationship) – are they higher or lower than expected?

Exam tips

You are not required to remember the exact formulae for statistical tests. If you need them they will be provided in the resources.

Patterns

These are often shown on maps. This also covers morphology (shape). Patterns can be:

- Nucleated or clustered together
- Linear
- Cuneiform or cruciform (cross-shaped)
- Regular
- Concentric
- Random or scattered/dispersed or amorphous.

Networks

In most forms of network analysis there are some key limitations:

- They treat all routes equally, regardless of their quality
- They focus on linkages rather than time or distance of journeys
- They look at planar (flat) networks – no flyovers, etc
- They ignore what uses that route.

Caution

When using statistical tests there are some key aspects you should always remember:

- To state your null hypothesis
- To state your alternative hypothesis (if null is disproved)
- To calculate the degrees of freedom
- To know the level of significance you are prepared to accept.

Remember

Statistical tests of association or relationship indicate the strength of that relationship but not the reasons for it. It is possible for there to be no causal link between the two variables that statistics suggest are linked.

Remember

If your analysis is incomplete or has gaps/flaws it is usual to conduct further research and revisit your methodology. This creates a feedback loop.

7.6 Present a summary of the findings and an evaluation of the investigation

Student book pages 293–95

Presenting your findings

Summary

- Draw a conclusion linked back to original question – proved or not?
- Why? Give main reason(s)
- Does it support the geographical model/concept? Suggest why or why not. Is there a flaw in your strategy/results or is the concept suspect?

Evaluation

- Discuss limitations – time, timing, equipment, method, personnel, location; explain effects of these on results
- Suggestions for improvement – other times/places, improved method (e.g. sampling methodology), more effective and/or robust equipment.

Exam tips

This stage is most likely to form the basis of a section B question so do ensure that you personalise it.

Limitations of the chosen model

It is rare that investigations do fully support the chosen model or concept. There are good reasons for this.

A) There are limitations in the models/concepts:

- Models/concepts are usually generalisations so they are unlikely to apply to your location or investigation
- They indicate what ought to happen in 'ideal' circumstances – often certain factors were held constant e.g. 'isotropic surface' (flat and featureless plain)
- Often they were formulated in a different country – usually but not exclusively the USA
- They were suggested some time ago, e.g. the Burgess model dates from 1925; a lot has changed since
- Many are not based on measurements in the field but are based on theory and logic.

B) There are limitations in the investigation in terms of the models/concepts:

- Different methodology, equipment, etc was used
- Different level of accuracy, timing, etc
- Insufficient sample size or number of measurements
- The investigation was only looking at one of the aspects of the model/concept or its surface expression.

Key words

Evaluation	Interpretation
Effectiveness	Conclusion

Anomalies

Sometimes your findings may show an anomaly in a model or concept. Your task is then to identify the particular circumstances that created that anomaly. This may lead to further geographical insights.

Limitations are suggested reasons why your investigation was not fully complete, accurate and reliable. It is rare for there not to be any. They can be grouped as:

- Initial planning limitations
- Data collection and recording limitations
- Data processing limitations
- Accuracy and reliability of the data.

Remember

If you identify a limitation you should always think of ways in which it could be overcome.

Often students say investigations have failed because they didn't prove a concept/model worked in their location. This is not a failure but you need to give some reasons why the model/concept was inappropriate in your area or location.

Improving your investigation

Improvements could include:

- Repeating the investigation in a different location
- Repeating the investigation at different times (day, week, year, season)
- Increasing the sample size
- Improving sampling strategy – e.g. unit, method
- Taking into account further variables
- Using better, more accurate and reliable equipment
- Finding further sources of data – especially secondary
- Re-formulating the title (the question being investigated)
- Making organisational improvements – better teamwork, etc.

Quick check questions

1. What is the name given to a distribution in which mean, mode and median coincide?
2. A standard deviation figure should always have what in front of it?
3. A total of 95 per cent of any data set should be within how many standard deviations of the mean?
4. What is a frequency distribution with two modes called?
5. What technique would you use to assess the pattern of the location of shops?
6. What would it indicate if the result of a Spearman's rank test was –1.6?
7. In a gravity model, what would be a better denominator than distance?
8. Suggest the chief reason why investigations often produce unsatisfactory results.

ExamCafé

Student book pages 248–99

This chapter represents one entire paper – Unit F764 or 20 per cent of the A level marks. It has two sections: A, in which you select one from three questions based on stages of the investigation process; and B, in which there is no choice but the two questions are based on stages of your own research or investigation.

Section A

Each question has three parts worth 5, 10 and 5 marks. The ordering of these parts may change but it will always be those mark divisions.

Sample question

a) Study figure 2, which shows a pie chart drawn up in an A level investigation to show numbers of pedestrians passing a point on the edge of a CBD as part of a survey of shopping patterns (n.b. figure not included).

i) Evaluate the effectiveness of Figure 2 in showing pedestrian flows into the CBD. [5 marks]

When looking at any figure that you are asked to evaluate, always consider:

- Is it appropriate for the type of data? In this case it was a pie chart, which wasn't appropriate
- Has it got the basics – title, scale, key?
- Has it been drawn correctly – use of colours, size, etc?
- What is the visual impact?

Even 5-mark questions are marked on a level basis. Generally, unless the question states otherwise, you can assume two points developed in depth would be sufficient to gain the higher level.

Student Answer

Figure 2 is a pie chart, which is not a good choice for this timed data. By converting the pedestrian count to percentages the reader is confused and cannot tell at a glance what is going on. The chart also lacks a title, with no indication where this is or when the survey was carried out.

This is a level 1 response, as evaluation is limited and little reference is made directly to the figure. Having said that, the student does try to evaluate and has made two clear points so it would be the top of level 1, gaining 3 marks.

If the next part is a ii) then there is still a link to the resource or figure.

Sample question

ii) Outline an alternative method of showing pedestrian movements into the CBD. Justify your choice of technique. [10 marks]

With this sort of question, it is important to get the balance right. 'Outline' means 'describe an alternative method' – such as flow lines, with a diagram to show you do understand the technique. But the main marks will go on your justification. Remember that 'justify' requires you to prove or show why (usually to provide evidence for your choice of one thing rather than something else).

The key word was 'movement' – hence the preference for flow lines over more static methods.

b) Assess the value of the use of photographs in an investigation. [5 marks]

Student Answer

It depends on the nature of the investigation (1). Photos are useful to locate the site of the investigation in terms of the nature of the area and time of year or day (2). They capture the instant (3) when the investigation took place. If well annotated (4) they give information about equipment used, techniques such as the type of sampling and can provide data (5) such as via a satellite photo or in an environmental quality survey. They are useful and valuable but need to be used sparingly and annotated carefully, as too many will turn the investigation into a photo album (6).

This is an effective answer because:

1. The candidate makes a good logical start
2. One clear use is stated
3. Justification here
4. Good geographical term
5. Two other uses of photos
6. There is a final cautionary note with some assessment.

To get full marks the assessment could have been sharper or there could have been a little more explanation of the uses of the various methods.

Section B

In section B you are required to answer two essays, which you are expected to base on your own investigation. You need to write down the title of the investigation.

Examples:

A shopping survey in Pershore town centre

Or

Does vegetation cover increase with distance away from the high water mark?

The first example tells the examiner little and seems more like a GCSE investigation. It lacks focus, unlike the second one. The latter creates an idea of the type of investigation in the mind of the examiner so they can see the relevance of the material in the answer.

Sample question

4) *Evaluate the method(s) you used to collect accurate and reliable data in your investigation.*

A clear evaluation is expected of data collection methods but the focus is on their reliability (dependability) and accuracy.

Student Answer

When we measured slope angles on the north and south facing slopes we found the clinometers were unreliable. Often one leg of the clinometer ended up perched on a stone so the angle was suspect and not all the clinometers read the same. We couldn't be sure that all the groups measured and recorded the angles in the same way we did.

This answer continued in a similar way. Reliability and accuracy were hinted at but not spelt out and the candidate rather missed the point of 'method', preferring to concentrate on the failings of the equipment used. Candidates must comment on the whole data collection method – i.e. sampling methodology etc, not just one limited aspect. This candidate did not get out of level 1. The answer lacked any of the exemplification that demonstrated a real investigation in the field.

It is worth noting that these essays are worth 20 marks and should take 30 minutes so they are not as demanding as those essays in the Global Issues (Unit F763). Introductions and structure are not as vital but an effective conclusion summarising the evaluation can be crucial. Communication is still credited but here it is the ability to draw out sensible and logical conclusions that tends to gain the marks in AO3.

Conclusion

Each chapter of this revision guide has given you an insight into the typical questions set on the topics, both as short answer questions and longer essays. Remember that writing well-structured essays in limited time, which answer the wording of the question set, is the vital A2 skill. This is particularly important if you want to gain an A*.

When you are ready to start your revision you need to devise a schedule that fits well with your lifestyle. It is best to build up your revision over time. Don't leave it to the last minute and try to cram it in.

Ideally, you should think in thirds. Divide the day into three 3-hour slots – morning, afternoon and evening – and initially revise during one of those slots (see schedule opposite). Which one you choose may depend on your other commitments, such as school/college or part-time job. The important thing is to do some revision each day but it doesn't always have to be in the same slot. This arrangement also gives you some free time at the weekend. Nearer to the exam, how long you revise for depends on how many subjects you are taking and how much you need to revise. At this stage, you should increase your revision to two slots a day. Schools usually start study leave a week or two before the exams. **Never** revise for all three slots, as you will get too exhausted to be able to analyse the questions carefully.

On the night before the exam look at your revision notes, which should by now be a set of mnemonics (e.g. PESP) and diagrams. Last-minute cramming never works. Remember, this exam is less a test of memory and more one of analysis and evaluation so try to have a refreshed mind.

Revision is vital but we all revise in slightly different ways. Find your most effective way and don't be swayed by what others do. It is **your** examination, not theirs. Remember the most common mistake is misreading the question. In addition you should make sure, in advance, that you have all the equipment you will need in the examination room.

There is also some final advice to be given. Most people get quite stressed before an exam, whether or not they appear tense. Try to relax by breathing deeply and walking about. Make sure you are there in plenty of time, as being even a little late can throw the most confident of candidates. What you do at the very end when you enter the exam room can be critical, so the following Exam Café gives some advice for those final minutes.

Sample grid for revision schedule

It is very useful to pin up your revision schedule as a constant reminder of your priorities. Fill in the boxes with what you plan to do. Notice how important lunch is. Your brain needs food and it is better to eat a good meal at lunchtime rather than in the evening, when it will send you to sleep.

	Morning 3 hrs	LUNCH	Afternoon 3 hrs	Evening 3 hrs
Monday				
Tuesday				
Wednesday				
Thursday				
Friday				
Saturday				

In the Exam Room

Remember these are longer exams than at AS, with fewer sections. You need to do well in **both** these papers if you are going to get an A*. Both are synoptic so you are expected to use material from your AS studies.

Essays are the main way of assessing your understanding and ability to evaluate and explain geographical concepts at A2 so a clear head is more important than one filled with endless detailed case studies. Unlike at GCSE, A2 examiners expect you to voice your own opinions, backed up with examples to justify your viewpoint. There are no right or wrong answers at A2.

Do:

- Read the questions carefully all the way through – each word is crucial
- Understand what the command words are and what they mean – they are similar to AS but there are some new ones, especially 'evaluate' and 'to what extent'
- Only read those sections that you have prepared for, otherwise you will be wasting time on sections you should not be doing
- Be careful, as in the Global Issues paper (F763) you may be able to do questions outside the units you have studied (be wary of this)
- Remember that the issue you identify for each of your answers in section A of F763 should be clearly shown in the resource(s) provided for that question
- Make a brief plan for each essay – it will be looked at if you don't finish the essay
- Keep an eye on the time – essays can run away with your time
- Leave time to read through your essay at the end to make sure it makes sense
- Structure your essays, with an introduction, middle paragraphs and conclusion
- Ensure you draw a conclusion that relates back to the question
- Answer your best question first – it should increase your confidence
- Personalise your essays in the Geographical Skills paper (F764) so it is clear what you did or contributed to the investigation
- Use detailed examples, even if not specifically asked to
- Write clearly, as your paper will be scanned for marking
- Use diagrams and maps where **appropriate**
- Locate things carefully
- Spell geographical terms and places correctly, e.g. Mississippi
- Remember to evaluate – recognise that the situation may vary with location, scale, time, level of development and from the viewpoints of differing groups in the community.

Don't:

- Spend too long in either paper on section A questions, as they carry fewer marks than the essays
- Suggest more than one issue in section A questions for the F763 paper – the marks are mainly given for suggesting appropriate solutions/management
- Spend 10 minutes on something worth 30 marks and 15 minutes on something worth 10 marks
- Use colours on diagrams, as they won't show up when scanned
- Panic. If you suddenly think you have done an answer incorrectly, stop and get on with another answer. You may find it wasn't wrong and when calmer you can go back to it
- Cross out work you think is wrong (until the end)
- Prepare possible answers before the exam. Essay titles are rarely ever repeated word for word.

Finally

Above all, read the question carefully, noting the key aspects and terms, and then answer it covering all these aspects, not giving the answer to the question you hoped would come up.

Answers to quick check questions

Chapter 1 Earth hazards

Mass movement events

1. Very wet conditions – adds weight and lubricates.
2. Soil creep. It is so slow that it poses no real hazard in the short term.
3. Water content is a lot higher in a flow.
4. It happened at a time when the school was full of children.
5. Often with explosives (mortar shells), although traditionally the alpine horn had a similar effect.
6. They need the wood for fuel.
7. Coastal cliffs, e.g. in the southern Isle of Wight.

Hazards and impacts of earthquakes and volcanoes

1. Destructive margins are the most hazardous because more energy is released at such margins.
2. Gas and oil pipes and electric wires rupture and domestic fires are overturned.
3. Already weakened buildings collapse while people are searching the rubble.
4. Because Kashmir is a very remote and little-known area.
5. The military is one of the few organisations with the personnel, resources and communication equipment needed to cope with an emergency.
6. Volcanoes usually already exist so people know they are there and are a potential hazard.
7. There were plenty of precursor events that warned people.
8. The ash and dust from volcanic eruptions often goes way up in the atmosphere and so falls many kilometres away and can even alter the climate in the short term.

Managing hazard events

1. People are unprepared, as they are asleep and in the dark. It is difficult to see escape routes etc. It's also psychologically more frightening at night.
2. There are more tall structures to fall and fire spreads more easily.
3. Sediments may suffer liquefaction. Solid rock cracks.
4. Africa has few earthquakes and few plate margins. It also has many seasonal climates with very wet seasons.
5. Tiltmeters indicate a swelling of the crust as magma rises in a volcano.
6. Oil. There is little evidence that it works.
7. They sprayed seawater on it to cool it.
8. A series of warning buoys has been set up to give earlier warning of a tsunami.

Chapter 2 Ecosystems and environments under threat

Components and changes in ecosystems and environments

1. Take-up of nutrients by plant roots.
2. There is a fall-off in energy and mass up the trophic pyramid. Not all energy and mass are fully converted when consumed by the level above.
3. The climate is warmer, with longer daylight hours leading to more energy input.
4. 75 times greater.
5. By deposition of material (e.g. a beach or lava flow) or by erosion of a surface (e.g. a wave-cut platform).
6. Humans work on a short timescale, at a local level and with a wide range of ways of impacting on succession.

Impact of human activities on ecosystems and environments

1. Every 20 minutes, although hundreds of new species are being discovered each year especially in remote rainforest areas, where a single tree may have several unique species – usually insects.
2. Ownership by the London Corporation from 1878 (pre-car).

3. Removing tree branches at head height as a 'crop'.
4. Only the rich can afford not to exploit the environment to the full.
5. High energy input, as they are on the equatorial belt of high insolation and high rainfall. Intense competition leads to diversity.
6. Wood is seen as sustainable energy and MEDCs need to diversify in order to reduce their dependence on fossil fuels.
7. High population growth so urgent need for food and money gained from transnational corporations (TNCs).

Benefits of conservation and protecting biodiversity

1. To maintain a healthy food chain so reducing pests that could damage their crops.
2. Salt-marshes act as a buffer against coastal flooding. Protecting them also counteracts the public perception of refineries as environmental polluters.
3. Few people go onto them (due to the risk of unexploded shells) so disturbance is minimal.
4. It is easier to control the specific use of the land.
5. Probably the fear that a reduction in biodiversity could lead to major diseases wiping out many of the species we rely on, e.g. wheat.
6. Ice caps will melt and the sea level will rise very fast, leading to larger waves and currents that will endanger the coral.

Chapter 3 Climatic hazards

Hurricanes and tornadoes

1. a) 5° N/S b) 35° N/S
2. The heat of the sea and the release of heat from the condensation of water increase the energy in a hurricane so it speeds up and there is little friction with the surface of the sea.
3. Flooding – from storm surge or heavy rain.
4. The Saffir-Simpson scale
5. The eye is still and calm so many people think the worst is over and come outside. They are therefore out in the open when the second eye wall passes overhead.
6. Because of the huge contrast in air pressure.
7. It occurred at 1.50 am and in November.
8. They are smaller and continue for shorter periods.

Hazards and impacts of atmospheric systems

1. Air is sinking so it warms up.
2. Clear skies allow heat to escape at night.
3. Due to rapid cooling at night, fog gets trapped by the inversion layer.
4. A cold front, as uplift is greater.
5. An occluded front.
6. The longer the sea crossing, the more moisture and warmth, leading to heavy snowfalls.
7. It kills plant cells.
8. They are relatively rare – about once every 20 years.

Managing climatic hazards

1. Trees stabilise the ground so reducing landslides. They also absorb water and act as a windbreak.
2. It is costly and not very successful.
3. People are reluctant to leave their homes.
4. Its sheer size (or you may view the failure of the levees as crucial).
5. There are benefits that outweigh their perception of the level of risk.
6. Because of the sheer cost and fear of panic. Also climatic hazards are often very unpredictable until the last minute.
7. Desertification, probably caused by overgrazing.

Chapter 4 Population and resources

Contrasts in population growth and policies

1. They gain work and their own independence so they can escape from being 'merely baby factories'.
2. Culture differs and they start at a higher birth rate.
3. a) People can now migrate out.
   b) People can now migrate in.
4. Changes in population growth power the desire to migrate.
5. Immigrants might vote a different way from the local population, they need more resources, and increased immigration often triggers a political backlash against the government.
6. By being used as a destination for convicts.
7. Most of sub-Saharan Africa.
8. It created a major sex imbalance, with far fewer girls.

Factors affecting supply and use of resources

1. There are often disputes about who owns or controls them, e.g. the cod wars between Iceland and the UK.
2. War puts great pressure on limited resources. In a life or death situation, cost is less important than survival and innovation becomes vital.
3. a) There are cheaper nylon substitutes. b) Nylon is based on fossil fuel resources (hydrocarbons), which will run out or increase in cost. Hemp is renewable.
4. It can lead to exploitation – oil producers forcing up prices when selling to the 'have nots' – or even wars.
5. Aluminium comes from bauxite, a very common residue of tropical weathering found in many tropical soils. It is almost ubiquitous.
6. Rain doesn't always fall where it is needed most. The water cycle may be a closed system but only on the global scale.
7. The price of tin has risen and it is increasingly being used in electronics.

Contrasting patterns of demand

1. Older people need more medical resources, both in volume and variety, for longer than younger people.
2. There are some very basic tertiary jobs that even very undeveloped cultures would have e.g. priest, soldier and seller.
3. Salt.
4. Higher incomes mean more consumer goods are demanded. Many of these contain steel, e.g. cars, TVs, fridges, etc.
5. People in LEDCs often need the animal to be alive to supply valuable products (e.g. milk, hair and dung) and they have a diet that is lower in protein and fat than people in the MEDCs.
6. Richer people have more water-using equipment (e.g. washing machines) and a lifestyle that encourages water-based recreation (e.g. swimming pools). Also, most water is used for power production.
7. 9.5 times.

Chapter 5 Globalisation

The problems and benefits of globalisation

1. To supply raw materials for their new industries.
2. It was designed in 1973 and the world wide web (www) started in 1989.
3. Some would say this is because the USA dominates the media and the Internet. Also, Chinese pictographs are difficult to fit on a keyboard.
4. Singapore.
5. They fear that their own culture will be swamped by western values.
6. The spread of Chinese culture and ownership (as in Africa).
7. Transnational corporations (TNCs).
8. The increasing pressure for political reform and democracy.

Aid and trade

1. Host market.
2. If one stage fails it all fails.
3. The country of origin – the USA.
4. In MEDCs – often their countries of origin. This is a very crucial and sensitive stage in production.
5. The export and import of services.
6. Manufactured goods.
7. With invisible trade, or if that isn't sufficient (as in the UK) then it's balanced by borrowing.
8. Travel and tourism
9. The USA has a trade embargo because it sees Cuba as an anti-USA communist state. It therefore uses trade as a weapon in its fight against communism.
10. Aid from one state directly to another.
11. They have small populations so aid per person looks huge and they often occupy strategic places in the world.
12. Saudi Arabia has vast sums of money from its oil revenue and it wishes to help fellow Muslims.

Chapter 6 Development and inequalities

Different levels of development

1. People and cultures can't agree on which factors contribute to the quality of life and how to measure them.
2. The faster pace of life and stress of growth may reduce people's happiness and quality of life.
3. Communist states, e.g. Cuba, North Korea.
4. Its annual economic growth rate of 10 per cent.
5. Bangladesh has such rapid population growth that most of its GDP growth is absorbed by the extra population. Little is left to invest in economic growth.
6. Ports and capitals.
7. Trickling down.
8. Tourism is relatively low cost in investment and the very remoteness of these areas is often their biggest attraction.

Social and environmental inequalities

1. Many will work informally or for themselves. It may also hide examples of underemployment.
2. 87 per cent.
3. The pressure tends to be at ground level so there is less pressure on tree-dwelling birds. Also, birds often adapt to using human structures better than mammals do.
4. Because of legal controls and the decline of manufacturing industry.
5. We are running out of sites and they often pollute the water table and generate methane (a major global warming gas).
6. Modern intensive farming causes a great deal of pollution.
7. 14 years longer in the suburbs.
8. Youthful – 22 per cent under 16 (shows impact of migrants with large young families).

Reducing social and economic inequalities

1. It's like a bank but it is owned and run by those who pay into it.
2. Those needing cheap housing, e.g. shantytown dwellers in LEDCs.
3. Puts them up.
4. 'Stick' policies punish you if you don't do something and 'carrot' policies reward you if you do. They are often used together.
5. You might see it as Myrdals or the multiplier model.
6. Currently, 80 per cent of air travel goes through the main London airports and airports create a lot of local growth.

Chapter 7 Geographical skills

Planning an investigation

1. They should be Specific, Measurable, Achievable, Realistic and Timed.
2. T – the Timing aspect.
3. On safety grounds – it's too risky.
4. Repetitive strain injury, eyesight problems and any of those that relate to accuracy of data collection.
5. The actual collecting or measuring of data.
6. It depends whether it is processed into a final form (secondary) or not (primary).
7. It immediately creates a bias (as not everyone has Internet access) and how do you obtain enough email addresses?

Presenting data

1. It shows the conditions (e.g. weather) at that moment in time.
2. Clear location, date/time and some way of showing scale.
3. The Earth's surface is curved so it won't fit on a flat page without distorting area, distances or directions.
4. 12 o'clock.
5. The smallest (unless you are following the order of other pie charts).
6. They will be scanned in black and white.
7. When you wish to show two variables on the same map/diagram or a positive versus negative pattern.
8. Interpolation.

Examples from the student book

1. It is unclear what constitutes a 'major' earthquake.
2. It shows data that is measured continuously. The average flow is rarely achieved – mode might be more informative.
3. This is one way of grouping data and there is some evidence that tornadoes occur in ten-year cycles but it can mislead, e.g. all of the 1940s killer tornadoes may have occurred in one month in one year.
4. Presumably this variable was not measured prior to that date. Actually the country did not exist until 1971!
5. It is not easy to see the exact amounts, as the range of values is so great. The smaller values especially need their exact value shown.

Presenting a summary and evaluation

1. Normal distribution.
2. + or –.
3. +/–2.
4. Bi-model.
5. Nearest neighbour.
6. That the test had been incorrectly done as answers must lie between +1 and –1.
7. Time (travel time).
8. The chief reason tends to be an inappropriate initial title or question for the investigation.